Carnivorous Plants
in the Wilderness

Carnivorous Plants
in the Wilderness

Makoto Honda

2024-April-26 b Hardcover

To my parents

In the wilderness, beyond the reach of civilization, unfolds nature's drama of survival.

CONTENTS

Preface **10**

Introduction **12**

1 Pitcher Plants *Sarracenia* **42**

2 Cobra Plant *Darlingtonia* **94**

3 Sundews *Drosera* **128**

4 Venus Flytrap *Dionaea* **166**

5 Butterworts *Pinguicula* **186**

6 Bladderworts *Utricularia* **222**

Distribution Map **264**

Glossary **266**

Bibliography **267**

Index **277**

Sarracenia rubra ssp. *wherryi*, in May, southern Alabama.

Preface

This is a book on the ecology of carnivorous plants, their lifestyle and surroundings. I created this book to take you into the wilderness. Or, perhaps, bring the wilderness right into your living room.

Like some of you, I became interested in carnivorous plants in my childhood. The idea of tiny plants devouring live insects intrigued me. I started to grow some. I found a group of people interested in these vegetable oddities — The Insectivorous Plant Society, founded in Japan in 1949.

In the United States, I visited many bogs and swamps over the span of decades in search of native carnivorous plants, many of which are endemic to this continent. During this time, I have seen large habitats being claimed for development.

As the human population increases, the health of the ecosystem around the world is becoming compromised and many species are being pushed to the edge of extinction. Carnivorous plants are no exception. Efforts are underway to preserve these unique species in nature but the time is running out for some.

This book is a photographic tribute to some of the most remarkable carnivorous plants in their natural environs that I was able to witness during my wanderings through the wildernesses of North America.

Thread-leaf sundew *Drosera filiformis* and a spider.

Introduction

The notion that some plants eat animals sounds strange. It is a deviation from our familiar concept of the food chain. Plants are eaten by herbivores and herbivores, in turn, are eaten by carnivores. Carnivorous plants have short-circuited the flow of energy within the ecosystem.

Historical records show there was strong reluctance on the part of eighteenth-century botanists to accept the idea that some plants had evolved to catch and digest animals for their nutritional needs. Charles Darwin was one of the first to demonstrate that some plants had indeed been adapted to the carnivorous habit. Subsequent studies using modern means have shown that the nutrients obtained from captured prey are absorbed through the trap leaf and are carried to the growth sites, suggesting that the plants do derive benefits.

The main requirements for the healthy growth of plants are sunlight, carbon dioxide, water, and some inorganic nutrients. A deficiency in any of these basic necessities creates a harsh environment for the plants. In an arid land, available water becomes the limiting factor. In a dense forest, competition for light is a life-or-death issue. In any adverse situations, the plant must adapt — or perish. A wide diversity of environmental conditions that prevailed throughout the evolutionary history of modern flowering plants has created a staggering array of properties found in the richness of the plant kingdom of our planet today.

There are places in the world where the soil is poor and plants cannot obtain enough nutrients through the root to sustain their growth. This particular environmental stress has given rise to a syndrome quite eccentric in view of a normal plant lifestyle. It is in such mineral-deficient environments found in some parts of the world that the plants that have adopted carnivory can be found.

Drosera capillaris with a bug.

CARNIVOROUS PLANTS ERAS

Charles Darwin was born in Shrewsbury, Shropshire, England, 1809. (Portrait at 22.)

" **A small society but worldwide presence.**

A relatively small group of these meat-hungry flowering plants have collectively come to be known as "insectivorous" or "carnivorous" plants based on their common ecology. There are over 750 different species of carnivorous plants recognized today, representing thirteen taxonomic families of angiosperm (flowering plant) classification. The geographical distribution of carnivorous plants encompasses the entire globe, extending over all continents except Antarctica. Some species grow widely throughout the world, while others are confined to a narrow geographical range.

CHARLES DARWIN PUBLISHED the book *On the Origin of Species* in 1859, at the age of 50, after decades of pondering the idea. Sixteen years later, *Insectivorous Plants* (1875) was published. The curious world of carnivorous plants did not escape the attention of the great mind after all. To this day, his work remains the most comprehensive classical account on the subject of carnivorous plants.

The book describes various carnivorous plants representing ten genera then considered carnivorous. Of note is his detailed observations of sundews (including their tentacle behavior) for which Darwin dedicated almost two thirds of his 400-plus-page treatise. Through a series of meticulous experiments, Darwin provided convincing evidence — in the air of surrounding skepticism at the time — that some plants truly had adopted carnivory to supplement their mineral shortcomings. The term "carnivorous plants" was also coined in the latter half of the nineteenth century, noting a wider range of animals the plants consumed.

Entering into the twentieth century, the year 1942 saw the publication of Francis E. Lloyd's *The Carnivorous Plants*. More species were added to the family of carnivorous plants, now counting fifteen genera (and carnivorous fungi). Lloyd spent one third of the book elucidating the complex triggering mechanism of bladderwort (*Utricularia*) traps. The modern concept of carnivorous plants was now being formed.

The latter half of the twentieth century is culminated by the advent of new technologies, including electron microscopes and radioactive isotopes. Various research has been conducted utilizing advanced tools. Newly acquired knowledge brought forward by modern science is now compiled in a monumental work by B. E. Juniper, R. J. Robins, and D. M. Joel in the book *The Carnivorous Plants* (1989). Peter Taylor published his taxonomic monograph *The Genus Utricularia* in the same year.

As we usher in the twenty-first century, molecular systematics continues to offer objective inferences of evolutionary relationships among flowering plants never before possible, lifting our understanding of carnivorous plants phylogeny to a new height.

LEFT: Poor choice of a landing site; a gnat sensing something's not right. *Pinguicula macroceras.* OPPOSITE: *Drosera rotundifolia* trying to secure a meal for the day. In May, northern California. The most profound energy transfer in the ecosystem takes the form of one organism consuming another. The organism that vanishes in the process is called the prey; the remaining, the predator.

DEFINITION

"You may submit your qualifications for membership consideration.

Carnivorous plants are green flowering plants. As such, they do carry out photosynthesis, like any ordinary green plants, but in addition they supplement their daily diet with animal protein.

The main nutrients selectively absorbed by carnivorous plants from animal prey are nitrogen (N) and phosphorus (P) but some other elements such as potassium (K: *kalium*) and magnesium (Mg) may be also utilized by some species.

The exact definition of carnivorous plants may differ among authors, and there is no consensus on the final count of species. But to be included in this exclusive rank of plant carnivores, a plant must minimally exhibit the ability to capture prey, digest it, and assimilate its nutrients. Many carnivorous plants also employ various attractions to lure prey. Last but not least, the plant must derive survival benefit from carnivory. The general description of carnivorous plants thus becomes:

1 Attraction of prey
2 Capture/retention of prey
3 Digestion of prey
4 Absorption of digested material
5 Derivation of benefit

Some botanists require plants to produce their own digestive enzymes. Others consider digestion aided by external organisms — be that bacteria, other microorganisms or commensals — valid and acceptable.

Lloyd (1942) rejected *Roridula*, South African herbs, because the plants' adhesive trap uses resin, not water-based mucilage. (The leaf cannot resorb resinous secretions.) Recent studies have revealed the existence of commensal bugs residing on the *Roridula* plants. The *Pameridea* bugs freely walk on the sticky leaf surface and suck nutrients from the trapped prey. The plant absorbs the bug's excretions deposited on the leaves.

The latest additions to the family of carnivores include a bromeliad species (*Brocchinia reducta* in the large Bromeliaceae family) having a primitive pitfall trap in the tightly formed rosette center. The inner leaf surface is coated with detachable waxy powder a microscopic structure of which has been shown to promote slippage of insects into the water. Trichomes (hairs) are found on the leaf that allow the absorption of digestion products. Along with the unusually yellowish leaf color (for a bromeliad) for prey attraction, this species was proposed (1984) to possess a collective syndrome sufficient for inclusion in the carnivorous family.

The list of carnivorous plants is likely to grow further.

The purple blossoms of *Pinguicula macroceras* in mid-April, northern California. Butterworts capture their prey with glandular hairs that cover the leaf surface. Typical prey are tiny winged insects such as gnats. Upon landing, victims are mired down in the sticky secretions and digested.

KINDS OF TRAPS

lthough traps come in a variety of colors, shapes and sizes among different species, the traps of all carnivorous plants are considered to be modified leaves in terms of their derivation. There are five basic trap types used by carnivorous plants.

Pitfall Trap The pitfall trap is the simplest of all trap structures found in carnivorous plants. Each single leaf grows into a pitcher that holds liquid at the bottom. In some primitive pitfall-trap carnivores, multiple leaves together form a water-retaining tank at the rosette base. The prey falls into the liquid and drowns. In some species, active enzyme secretions are detected, but in many the digestion is heavily aided by externally introduced microorganisms.

" All carnivorous plant traps are considered to be modified leaves.

Typically, the pitchers are colorfully decorated and are marked with ultraviolet absorption patterns. Nectar is often offered as an attractant along with a fragrance in some. The pitfall trap is considered passive because it does not have any moving parts.

This trap type is found in five families, a total of eight genera of carnivorous plants: some 150 species of tropical pitcher plants in the Old World (*Nepenthes* in the family Nepenthaceae), three genera of the New World pitcher plants totaling some 30 species (*Sarracenia, Darlingtonia* and *Heliamphora* in the family Sarraceniaceae), the Western Australian pitcher plant (*Cephalotus* in the family Cephalotaceae), and recently recognized carnivorous monocots (*Brocchinia* and *Catopsis* in the family Bromeliaceae and *Paepalanthus* in the family Eriocaulaceae).

Adhesive Trap Some carnivorous plants cover their leaves with fine hairs tipped with a sticky glue. This is called an adhesive or flypaper trap. In the sun, a drop of mucilage glistens like a dewdrop. Insects are known to be attracted to shiny blobs. Many species in this group have also developed sensitivity to physical as well as chemical stimulation, and enzyme secretions are detected in many species. Bending of the mucilage-holding hairs (tentacles) as well as leaf folding is also seen in some species.

The adhesive trap is found in eight families, a total of nine genera. The largest in number is sundews (*Drosera* in the family Droseraceae) with about 240 species, followed by butterworts (*Pinguicula* in the family Lentribulariaceae) containing some 90 species. Both sundews and butterworts have a worldwide distribution. The remaining genera of this group are the Portuguese dewy pine (*Drosophyllum* in the family Drosophyllaceae), a West African tropical liana (*Triphyophyllum* in the family Dioncophyllaceae), several species of the rainbow plants from Australia (*Byblis* in the family Byblidaceae), two species of *Roridula* (family Roridulaceae), some Devil's claw plants (one species of *Ibicella* and two species of *Proboscidea*, both in the family Martyniaceae), and a handful of newly added Brazilian species of *Philcoxia* (in the family Plantaginaceae).

Snap-trap We have come to the world-famous Venus flytrap (*Dionaea* in the family Droseraceae) which employs a snap-trap, sometimes referred to as a steel trap or bear trap. The trap has a terminal leaf blade that is divided into two half-shell-shaped lobes connected along the midrib. This trap has developed highly specialized sensory hairs on the inner lobe surface that initiate trap closure when stimulated. Along with the swift movement of trap leaves, the snap-trap may be considered the triumph of plant carnivory. The natural distribution of the Venus flytrap is confined to the coastal savanna of North and South Carolina in the United States.

This trap mechanism is shared only by one other species in the Old World, the waterwheel plant (*Aldrovanda* in the family Droseraceae), an aquatic cousin of the Venus flytrap. *Aldrovanda* occurs in Europe, Africa, Asia and Australia. Interestingly, the natural distribution of the waterwheel plant does not extend over into the New World where the Venus flytrap grows.

Suction Trap Some aquatic and semiaquatic species have developed a mechanism that sucks up prey into a pouch-like trap. Well over 200 species of bladderworts (*Utricularia* in the family Lentibulariaceae) bear numerous tiny sacs, or bladders, in the water and in the water-logged soil to capture small aquatic creatures. The prey is sucked into the trap in the blink of an eye upon triggering. The trap also self-resets in a matter of half an hour. The sophistication — as well as mechanical subtleties — of this trap is without parallel in the plant kingdom. In spite of its apparent sensitivity, the trap-triggering action itself is considered purely mechanical. The bladderworts have a worldwide distribution.

Lobster-Pot Trap In a closely related genus of corkscrew plants (*Genlisea* in the family Lentibulariaceae), there are nearly 30 species of semiaquatic plants occurring in South Africa and South America. In addition to their regular green leaves above ground, these plants grow white, Y-shaped (upside down) slender leaves underground, forming unique, spiral lobster-pot traps that capture small aquatic prey.

TOP LEFT: Pitfall traps of a pitcher plant (*Sarracenia purpurea* ssp. *purpurea*) in northern Michigan, in July. The trap retains liquid at the bottom. The nectar-secreting walls are very slippery that conduct unwary insects into the depth of the pitcher. TOP RIGHT: A sundew leaf (*Drosera rotundifolia*) holding a tiny spider. In northern California, in May. The leaf is covered with tentacles that are tipped with adhesive mucilage. The tentacles bend during prey capture. The whole leaf may also fold to secure the prey. BOTTOM LEFT: Bladderwort (*Utricularia*) traps in the water, that capture small aquatic prey with enormous suction force. The tiny hairs growing on the trap door function as the deadly trigger for the trap. BOTTOM RIGHT: The Venus flytrap (*Dionaea muscipula*) with its red-tinted, nectar-baited lobes wide open. Each lobe has three sensory hairs that detect the presence of prey in the trap. In July, North Carolina.

Cobra plants (*Darlingtonia californica*) in
high noon. In August, northern California.

AN ASSEMBLAGE OF
COMMON ELEMENTS

" Traps have commandeered all the alluring elements of real flowers.

Various trapping mechanisms developed by carnivorous plants may, at first glance, appear unworldly. It is hard to imagine that a small group of these meat-eating plants have come up with such clever and sophisticated trap designs, not to mention uncanny deceit.

If we analyze more carefully, however, individual components used to achieve these mechanisms are something quite common throughout the plant kingdom.

We see a parallel between the alluring strategies of flowers and traps. Every element used by insect-pollinated flowers — color, ultraviolet absorption, fragrance and nectar — is exploited by the traps as well. Various motions in plants are not uncommon either. Plants lean toward light. Flowers open by flexing their petals in response to light and warmth. Vines curl around the tree branch to cling, all due to differential cell growth or turgor change.

There are many non-carnivorous plants that produce viscous substances to protect their flowers against pests. Carnivorous plants have taken the idea one step further. Normal plant leaves are capable of absorbing fertilizer from the leaf surface, as seen in widely practiced foliage spraying. So the absorption of nutrients in the trap — as in the animals' stomach — is nothing out of the ordinary.

In carnivorous plants, what is remarkable is the degree to which many of these features are assembled together to achieve a specific end.

Attraction — a non-fatal variety. Pink flowers of *Calopogon tuberosus,* an orchid occurring widely over the eastern half of North America, often found in acidic bogs in southeastern pine savannas and marl fens in the north where many carnivorous plants grow. In mid-July, northern Michigan.

ABOVE: **Fatal attraction.** The Venus flytrap uses bright colors and sweet nectar to attract visitors — as do many flowers — but with a differing intent.

LEFT: **Nectar, anyone?** New spring pitchers of *Sarracenia leucophylla* covering the longleaf pine savanna in southern Alabama, in early May. Colorful pitcher leaves are often mistaken for flowers by flying insects and uninitiated human observers alike. RIGHT: *Drosera anglica* growing in a southern Oregon swamp. In early July.

Sarracenia purpurea ssp. *purpurea* growing in a "kettlehole bog" in central Wisconsin, in July.

POLLINATOR - PREY DILEMMA

" How not to eat pollinators ...

It is generally assumed that most carnivorous plants are insect-pollinated. We have noted that many traps mimic real flowers in their attraction scheme, implying that traps aim to capture "pollinators." This leads to an apparent paradox: The plants need to consume prey for nourishment and at the same time they need live pollinators for successful transport of pollen for reproduction.

Habitat Separation In some aquatic/semiaquatic species, this problem is averted by the clear separation of prey-trapping and pollination spheres. In *Utricularia, Genlisea,* and *Aldrovanda,* the trap devices that capture small aquatic animals lie in the water or in the damp soil. Meanwhile their flowers protrude in the air on tall peduncles that get pollinated by flying insects. For terrestrial carnivorous plants, prey and pollinators share the same habitat. How do they manage to reconcile this seeming dilemma?

Separation in Time Timing is one way in which pollination and trapping occur in sequence. In many eastern North American pitcher plants, *Sarracenia,* inflorescence occurs a few weeks prior to the production of new pitcher leaves of the season. This ensures that, during the anthesis, there are few functional pitchers to trap visitors to the flower. The Californian pitcher plant, *Darlingtonia,* follows suit. In some butterworts, *Pinguicula,* that form hibernacula (winter buds), a flower stalk emerges from the rosette center before active glandular leaves unfold.

Separation in Space Many carnivorous plants physically separate trapping and pollinating zones. The Western Australian pitcher plant, *Cephalotus,* produces an unusually tall flower stalk that bears small white flowers. The majority of prey for *Cephalotus* pitchers (typically 2-4 cm in size) are crawling insects, mainly ants. The flowers borne on a tall scape (that sometimes reaches 60 cm in height in the wild) are well isolated from the trapping zone on the ground below. A similar observation is made in Venus flytrap as well as many rosette sundews, if to a lesser degree in some species.

Insect Segregation This means different classes of insects are attracted to traps and flowers. Sometimes this is done by differing attraction schemes, be that color or other enticements, sometimes by size differentiation. In the American Southeast, many temperate butterworts bloom in early March. Their rosette glandular leaves are already active, catching tiny winged insects such as gnats. However, large, conspicuous flowers attract bees and other insects which are far too powerful for the glandular leaves to capture. Some African and Australian sundews also produce relatively large flowers seemingly intended for larger insects than the typical prey commonly observed trapped. Similarly, the Portuguese dewy pine, *Drosophyllum,* produces large, bright yellow flowers which create a sharp visual contrast to its glandular trap leaves.

TOP: Cobra plant blossoms in northern California, in May. The dangling flowers are borne on a tall scape for spatial separation. Flowering also occurs prior to new pitchers, thus providing temporal separation between pollination and trapping. OPPOSITE BOTTOM LEFT: Venus flytrap flowers bloom on a tall stalk, achieving spatial separation from the traps on the ground. OPPOSITE BOTTOM RIGHT: Flowers of a pitcher plant hybrid. New pitchers are yet to emerge, which may offer a hint of parentage. In the Okefenokee Swamp, Georgia, in May.

ABOVE: Is this guy a real pollinator or just basking? The bright yellow blossom of the yellow butterwort (*Pinguicula lutea*) often attracts bees to their flowers, though a bee is way too large to be trapped by the adhesive leaves. Typical prey for the butterworts are tiny gnats. In early March, in the Florida Panhandle.

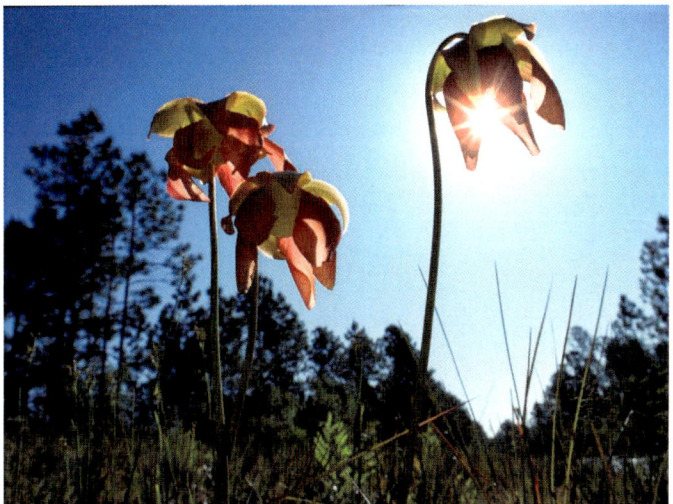

27

ECOLOGY

Carnivorous plants grow in mineral-deficient soils in various types of wetlands. In bogs and swamps, the accumulated organic matters (mainly dead plants) are not readily decomposed due to the high acidity of the water, and the minerals contained within are not released into the soil. This creates a nutritionally challenging environment for the plant occupants. Fen habitats in the north are also known to be nutrient-poor for the majority of nutrients are retained within the water-saturated developing peat layers.

Plant growth is limited by availability of the required nutrients. Nitrogen and phosphorus are major plant nutrients. Limited availability of necessary minerals keeps nutrient-hungry and often aggressive species in check. This creates rich fen vegetation. Fens are vulnerable to a slightest change in nutrient regime that causes increases in the limiting nutrient.

Pioneers Carnivorous plants can preferentially survive — and often flourish — in these inhospitable conditions because of their special adaptations that enable them to supplement the limiting nutrient from a different source. Carnivorous plants, in fact, are often seen as pioneers in newly developed frontiers that are too harsh for plants with a conventional lifestyle.

Weak Competitors Acquisition of the carnivorous trait did not come without a price, however. Carnivorous plants, in general, are weakly rooted, because one main function of the root — absorbing nutrients — is delegated to their trap structures. The majority of carnivorous plants are susceptible to desiccation and shading, often exhibiting a low tolerance for competition.

When a wet savanna in North America is left without periodic fires, the habitat undergoes a gradual transformation into an advanced forest. As the land becomes drier, the chemical compounds stored in organic matter get decomposed into plant-accessible forms (mineralization), that makes the soil richer. This is but a fate of wetlands in the natural process of

> " **Carnivorous plants can preferentially survive in a harsh environment because of their adaptations.**

transition. More plants start to grow, robbing carnivorous plants of precious sunlight. Carnivorous plants are known to be the first to disappear from the scene, succumbing to competitive invasion from other species. The environmental niche they enjoyed has all but ceased.

Cost Analysis Carnivory incurs a heavy cost for (a) trap construction: one-time expenditure, often achieved at the expense of photosynthetic capacity, and (b) trap operation: recurring cost during the life of the trap, that includes production of attractants, energy needed for trap-triggering and prey digestion. The overall cost varies largely depending on the type of traps used.

Maximizing Return The expenditure must be recouped somehow. It is not surprising that the benefit of carnivory is more readily manifested under the condition where the negative factor is largely confined to mineral deficiency. This is an environment which is extremely poor in soil nutrients, yet near-optimal in other aspects, having sufficient sunlight, enough water and so on. Here, nutritional supplements through carnivory directly improve the overall outcome without reaching a benefit plateau too quickly.

With heavy investment, carnivorous plants must balance the check book carefully. In nature, the margin of error is thin. This explains why carnivorous plants are only found in a narrow environmental niche.

Exceptions There are some exceptions in the carnivorous plant world. Portuguese dewy pine (*Drosophyllum*), some Australian rainbow plants (*Byblis*) and the devil's claws (*Ibicella* and *Proboscidea*) grow in dry soil and all have a substantial root system.

LEFT: A spectacular *Darlingtonia* colony formed on a mountain slope in southwestern Oregon, in May. OPPOSITE: *Sarracenia* hybrid (*S. flava* x *leucophylla*) in southern Alabama, in May.

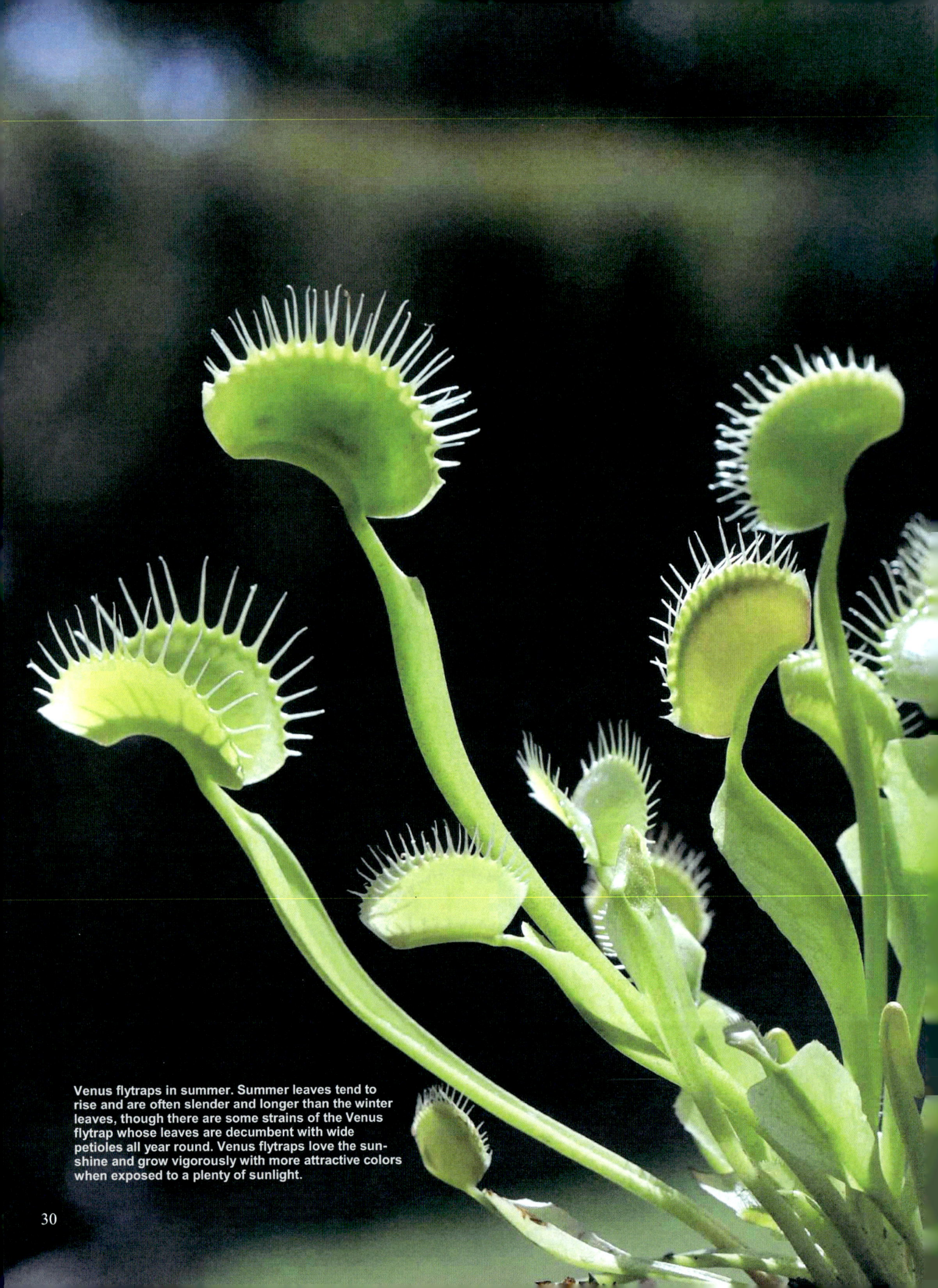

Venus flytraps in summer. Summer leaves tend to rise and are often slender and longer than the winter leaves, though there are some strains of the Venus flytrap whose leaves are decumbent with wide petioles all year round. Venus flytraps love the sunshine and grow vigorously with more attractive colors when exposed to a plenty of sunlight.

" The peril of man-eating trees in our backyard.

If we accept that the carnivorous habit is beneficial to the plants and does increase their chance of survival, we must beg the ultimate question: Why haven't many more members of modern flowering plants adopted carnivory in this animal-rich planet? Well over 250,000 species of flowering plants exist on Earth, yet there are only 750 or so carnivorous plants. This is a miniscule one quarter of one percent of the total species of angiosperms.

We also make note of the fact that the majority of carnivorous plants are small herbaceous plants. *Triphyophyllum peltatum* is a tropical liana from West Africa (recognized as a carnivore in 1979) that is said to grow to a height of ten meters (30 feet). However, the plant abandons the carnivorous lifestyle as it reaches a certain size. This plant produces glandular carnivorous leaves only during the juvenile stage (less than 50 cm tall), just before the rainy season, possibly to supplement nutrition in anticipation for flowering.

Is plant carnivory such a high-risk venture in light of the cost/benefit consideration?

I STOOD IN FRONT OF a tall eucalyptus tree. The eucalyptus is a leafy tree. In my mind's eye, I replaced all the foliage with sundew leaves covered with glistening, mucilage-tipped tentacles. In summer, this tree would capture hundreds of mosquitoes every evening!

With the thick, woody trunk, however, the amount of catch as measured in a biomass ratio is paltry — a tiny sundew plant that has managed to capture one fat mosquito would fare much better. To put it another way, a bigger stomach demands a more substantial meal. To make a meaningful nutritional contribution, this carnivorous tree would have to shoot for something more than a swarm of mosquitoes, say, a wild raccoon or two during the growing season. But to construct such a trapping apparatus out of the leaf structure would be a prohibitively costly proposition.

Nature's evolutionary journey thus far has not taken us to a world where our mailman is likely to encounter the sign that reads: "Beware of man-eating trees out of hibernation."

TOP: A crane fly falls victim to the voracious appetite of sundews. *Drosera rotundifolia*, in May, northern California. BOTTOM: Coniferous forest surrounding a *Darlingtonia* bog in northern California. Rest assured these trees are *not* man-eating. Lean against them and enjoy the fresh mountain breeze.

Purposeful transmutation. "If you come close
enough, I *will* eat you." *Darlingtonia californica*.

A blinding explosion of yellow blossoms (*Utricularia cornuta*) in a marl fen along the shore of Lake Huron. In northern Michigan, mid-July.

CLASSIFICATION

G eneral criteria for classification of flowering plants include morphology (of reproductive organs), phytochemistry (chemical compounds in plants) and biogeography (geographic origin and distribution). A traditional carnivorous plant classification is based on morphological analysis of floral as well as vegetative structures of the plants, including their trap devices.

Phylogenetic systematics attempts to construct a classification of organisms that reflects descent. Molecular biology and genetics research in the past decades have helped shed a tremendous amount of light on our understanding of evolutionary relationships among flowering plants, providing quantitative data for objective inferences of angiosperm phylogeny.

Molecular systematics relies on DNA sequencing that determines the exact order of nucleotides in a given DNA segment. Coupled with the modern computing power for subsequent analyses utilizing the cladistic approach, interrelatedness among targeted organisms based on their genetic sequences is probabilistically inferred, largely eliminating room for subjective interpretations that have resulted in differing taxonomic views in the past.

Radically revising the traditionally accepted classifications, all twenty-one currently recognized carnivorous plant genera are now placed in five orders, strongly suggesting that the transition from non-carnivore to carnivore had taken place in multiple lineages of flowering plants.

Poales The order Poales (belonging to monocots) contains carnivorous plants possessing a primitive pitfall trap in the family Bromeliadiceae (*Brocchinia* and *Catopsis*) and the family Eriocaulaceae (*Paepalenthus*). A water-retaining tank is formed by overlapping leaves in the rosette center. The leaves surrounding the tank are coated with loose, easily detachable waxy powder that causes an insect to lose its foothold. These taxa are sometimes only regarded as pseudo-carnivorous or quasi-carnivorous because of the primitive nature of their traps.

Oxalidales *Cephalotus* (in the family Cephalotaceae) is placed in the order Oxalidales. This is a loner and there is no other carnivores in this order. In fact, *Cephalotus* is the only carnivorous plant found in the "rosids," one of two major clades of flowering plants in the "eudicots" (see Phylogenetic tree, p. 41). This Australian pitcher plant is a unique plant occurring in a small area in Western Australia. The plant produces two kinds of leaves; a normal, non-carnivorous leaf and a pitcher-shaped pitfall trap. Its pitcher strikingly resembles that of certain tropical pitcher plants *Nepenthes*, but the *Cephalotus* pitcher evolved independently and is not related to other pitfall traps. The similarity is a result of convergent evolution — acquisition of characters resembling each other in function due to similar environmental conditions, not due to common ancestry.

Caryophyllales The order Caryophyllales is teemed with

carnivores: Droseraceae (*Drosera*, *Aldrovanda*, and *Dionaea*), Nepenthaceae (*Nepenthes*), Drosophyllaceae (*Drosophyllum*) and Dioncophyllaceae (*Triphyophyllum*). A traditionally accepted close relationship among *Drosera*, *Dionaea* and *Aldrovanda* based primarily on floral and pollen morphology has been favorably supported in molecular systematics.

Dionaea and *Aldrovanda* both possess a snap-trap, the most advanced prey trapping mechanism to be found among carnivorous plants. Although *Dionaea* is a land plant while *Aldrovanda* is aquatic, they share a great deal of common features in their snap-trap mechanism. Molecular analyses show the two species are sister to each other, suggesting a common ancestral origin of these snap-traps. Furthermore, *Dionaea* and *Aldrovanda* form a clade which is sister to *Drosera*. This strongly supports a speculation that the common ancestor of *Dionaea* and *Aldrovanda* is a sundew-like plant.

Nepenthes is a very diverse genus, containing over 100 species of tropical pitcher plants mainly from Southeast Asia. The plants develop a pitcher at the tip of a long tendril extending from the midrib of a regular-shaped leaf. *Nepenthes* and *Drosera* are speculated to have been derived from a common ancestor. The branch leading to *Nepenthes* has further developed into *Drosophyllum* with glandular leaves, like *Drosera*.

The carnivorous feature then seems to disappear and stay dormant for a while, only to re-emerge in Dioncophyllaceae: The monotypic genus *Triphyophyllum* is a tropical liana, somewhat reminiscent of *Nepenthes*, but with slender, adhesive glandular leaves, quite similar to those of *Drosophyllum*. In *Triphyophyllum*, these carnivorous leaves are produced only seasonally, and only in the young plant before it enters the adult phase of a climbing liana.

The adhesive tentacles of *Drosera*, *Drosophyllum* and *Triphyophyllum* are all multi-cellular in structure, possess a xylem in the stalk center, and are measurably more complex than a typical hair commonly found in plants.

It is as though, with the acquisition of an adhesive-trap trait, the branch leading to the common ancestor of *Drosera* and *Nepenthes* had been predisposed to carnivory.

Ericales DNA sequence comparison supports placement of Sarraceniaceae and Roridulaceae in the order Ericales, corroborating the traditional grouping of the two on the basis of phytochemical similarity.

The New World pitcher plant family Sarraceniaceae contains three genera, *Sarracenia*, *Darlingtonia*, and *Heliamphora*. In terms of their trap leaf formation, all these pitchers are rolled leaves with opposite margins fused together, quite different from pitchers of *Nepenthes* and *Cephalotus*.

Within the family Sarraceniaceae, *Darlingtonia* is found to be sister to *Heliamphora* and *Sarracenia*, suggesting that *Darlingtonia* diverged early in the evolution of the family, long before the separation of *Heliamphora* and *Sarracenia*.

Roridula has developed sticky leaves with powerful glue of resin. The plant resembles in shape some species of *Drosera*, but the similarity is more apparent than real. *Roridula's* adhesive trap fundamentally differs from other flypaper traps in

Classification of Carnivorous Plants

ORDER Poales

FAMILY **Bromeliaceae** (57 genera / 3170 species) Carnivorous Bromeliads
GENUS *Brocchinia* (2) B. hectioides B. reducta
GENUS *Catopsis* (1) C. berteroniana
FAMILY **Eriocaulaceae** (10 genera / 1200 species)
GENUS *Paepalanthus* (1) P. bromelioides

ORDER Oxalidales

FAMILY **Cephalotaceae**
GENUS *Cephalotus* (1) C. follicularis Western Australian Pitcher Plant

ORDER Caryophyllales

FAMILY **Droseraceae**
GENUS *Drosera* (244) Sundews
GENUS *Aldrovanda* (1) A. vesiculosa Waterwheel Plant
GENUS *Dionaea* (1) D. muscipula Venus Flytrap
FAMILY **Nepenthaceae**
GENUS *Nepenthes* (160) Tropical Pitcher Plants
FAMILY **Drosophyllaceae**
GENUS *Drosophyllum* (1) D. lusitanicum Portuguese Dewy Pine
FAMILY **Dioncophyllaceae** (3 genera / 3 species)
GENUS *Triphyophyllum* (1) T. peltatum West African Carnivorous Liana

ORDER Ericales

FAMILY **Roridulaceae**
GENUS *Roridula* (2) R. dentata R. gorgonias
FAMILY **Sarraceniaceae**
GENUS *Sarracenia* (8) Eastern North American Pitcher Plants
GENUS *Darlingtonia* (1) D. californica California Pitcher Plant / Cobra Plant
GENUS *Heliamphora* (23) Marsh Pitcher Plants

ORDER Lamiales

FAMILY **Plantaginaceae** (19 genera / 1700 species)
GENUS *Philcoxia* (7)
FAMILY **Byblidaceae**
GENUS *Byblis* (8) Rainbow Plants
FAMILY **Martyniaceae** (5 genera) Devil's Claw
GENUS *Ibicella* (1) I. lutea
GENUS *Proboscidea* (2) P. lousianica P. parviflora
FAMILY **Lentibulariaceae**
GENUS *Pinguicula* (95) Butterworts
GENUS *Genlisea* (30) Corkscrew Plants
GENUS *Utricularia* (247) Bladderworts

Lamiales In evolutionary terms, Lamiales is a young angiosperm order. Phylogenetic analyses indicate that in the order Lamiales the transition to carnivory had taken place independently in multiple families — Lentibulariaceae, Martyniaceae, Byblidaceae and Plantaginaceae.

In spite of the distinctly different trapping mechanisms found in the three genera of the family, Lentibulariaceae forms a clearly defined taxonomic group primarily based on floral morphology as well as phytochemistry (presence of acteoside). The basal genus in the family, *Pinguicula*, has developed an adhesive trap. The surface of the butterwort's leaf is covered in numerous hairs tipped with mucilaginous secretions. These hairs are a simple, single-celled hair supporting multi-cellular glands at the tip that produce mucilage. The other two genera, *Genlisea* and *Utricularia,* are closely related, both possessing a unique and complex underwater trap to capture tiny water animals. *Utricularia* appears to be more derived of the two. *Pinguicula*'s sessile glands are considered homologous to the quadrifid glands in the *Utricularia* trap.

Molecular systematics places Byblidaceae (*Byblis*) and Martyniaceae (*Ibicella*, *Proboscidea*) in the same order Lamiales. These have all developed an adhesive

that the secretions are resinous, distinct from water-based, mucilaginous adhesive traps found in other carnivorous species.

Although their trapping mechanisms are distinctly different between Sarraceniaceae and Roridulaceae, both absorb the nutrients by a very similar mechanism of cuticular discontinuity on the leaf surface.

trap similar to *Pinguicula*. The hairs of *Pinguicula*, *Byblis* and *Ibicella* (having a multi-cellular stalk) are a simple hair, commonly found in many other plants, and do not possess the complexity of sundew tentacles. *Byblis'* stalked glands are very similar to those of *Pinguicula*.

All carnivorous plants in this order produce zygomorphic (bilaterally symmetric) flowers.

Exuberant blossoms of a rare butterwort,
Pinguicula ionantha, an endemic of the
Florida Panhandle. In early March.

EVOLUTION

Nothing provides us with a more direct way of seeing the past than well-preserved fossil records. Unfortunately, the fragile structure of carnivorous plants does not lend itself well to fossil formation. We have only a handful of fossil records — mainly limited to fossilized pollen and seeds. Recently, there has been a discovery of a fossilized leaf suspected to be a precursor of the modern pitfall trap. From what we can deduce from these, many carnivorous families were already well established from the beginning of the Paleogene period (some 66 million years ago).

Pitfall-Trap Adaptation From the viewpoint of evolutionary development, the pitfall appears the simplest scheme for prey capture. Suppose a small amount of rainwater remains on a tiny dip on the leaf surface. An insect drowns in it, and eventually decays due to bacteria. The decomposed materials enter the leaf (absorption) — the plant receives unexpected nourishment. This unworthy event may provide survival benefit in a mineral-deficient environment, and a plant with a deeper, more

> ## " Phylogenetic systematics attempts to construct a classification that reflects descent — a family tree.

efficient pitfall may preferentially survive to carry the useful trait to the next generation.

The New World pitcher plants, the tropical pitcher plants and the Western Australian pitcher plant all independently developed pitfall-trap leaves in one form or another to exploit the opportunities. Fossil of a leaf of *Archaeamphora longicervia* from northeastern China, dating back to early Cretaceous (possibly 100 million years ago), shows superficial similarity to the extant New World pitcher plants, but its affinity to the Sarraceniaceae family is doubtful.

Adhesive-Trap Adaptation There are many non-carnivorous plants known that capture bugs with sticky glues. This is a defense mechanism against pests, particularly to protect their reproductive organs, or flowers. The fact that many adhesive-trap carnivores today (and their relatives) retain glandular hairs on their calyx and flower stems is suggestive of the evolutionary path they had taken. As in the pitfall scenario, the trapped insect may decay and the nutrients absorbed through the contact surface. This accidental benefit may push the adaptation toward a more efficient trap/absorption mechanism if the selective pressure of the environment so warrants.

Whether or not non-carnivorous adhesive traps exist today that are closely allied to the present-day carnivores, this scenario seems plausible in its development sequence. Given what we do not know, the transition from sticky defense to adhesive

carnivores seems a comparatively small step where nutritional supplement proves advantageous for survival.

There are eight carnivorous families that developed an adhesive trap. Of these, three families are closely related, so there appears at least six independent occurrences of adhesive -trap adaptation.

Snap-Trap Adaptation Only two snap-traps exist: *Dionaea* (Venus flytrap) and *Aldrovanda* (waterwheel plant). Molecular phylogenetics supports the idea of a common ancestral origin of the two. Molecular evidence further suggests that the common ancestor of *Dionaea* and *Aldrovanda* is descended from an ancient sundew-like plant.

Sundews' tentacles behave differently depending on where they grow. The marginal tentacles around the perimeter of the leaf blade are the longest. When directly stimulated, the tentacles bend toward the center of the blade. The short central tentacles, when stimulated, do not move, but transmit stimuli to other tentacles. The Venus flytrap's stiff marginal teeth around the trap lobes are observed to possess a xylem in the center, as do sundew tentacles. The leaf blade venations of *Drosera* and *Dionaea* are also shown to be very similar. It is possible that the marginal teeth of the Venus flytrap correspond to the marginal tentacles of sundews, while the sensitive trigger hairs came from sundew's central tentacles.

The *Drosera* tentacles usually move slowly, but in some species they react at an astonishing speed. The "snap" tentacles of *D. burmannii* will bend toward the leaf center in a matter of seconds after stimulation. In the case of *D. glanduligera*, the speed is measured in a fraction of a second, reaching or surpassing a record held by the Venus flytrap.

This is intriguing in view of a possible evolutionary scenario in which the snap-trap had been derived from an ancient adhesive trap. Without any fossil records revealing a hint of intermediate forms, how and when this transition unfolded remains unknown. Evidence indicates active speciation of "*Aldrovanda*" commencing around the end of the Eocene epoch some 40 million years ago — as revealed by fossils mainly of pollen, some seeds, and possibly one leaf — though only one species of the waterwheel plant survived to the present day. Whether the snap-trap transition took place underwater or above, a huge, quantum leap of mutation is likely to have played a role in the creation of a snap-trap. The event, it could very well be argued, is more astonishing and rarer in the development of plant carnivory, far more serendipitous than a mere emergence of pitfall or adhesive carnivores out of non-carnivores. Only one snap-trap event took place, as far as we know, in the history of modern flowering plants, that gave birth to the progenitor of our beloved Venus flytrap.

Suction-Trap Adaptation It is difficult to figure out how the suction trap came about. The complexity of its triggering mechanism mesmerizes researchers to this day. Only one

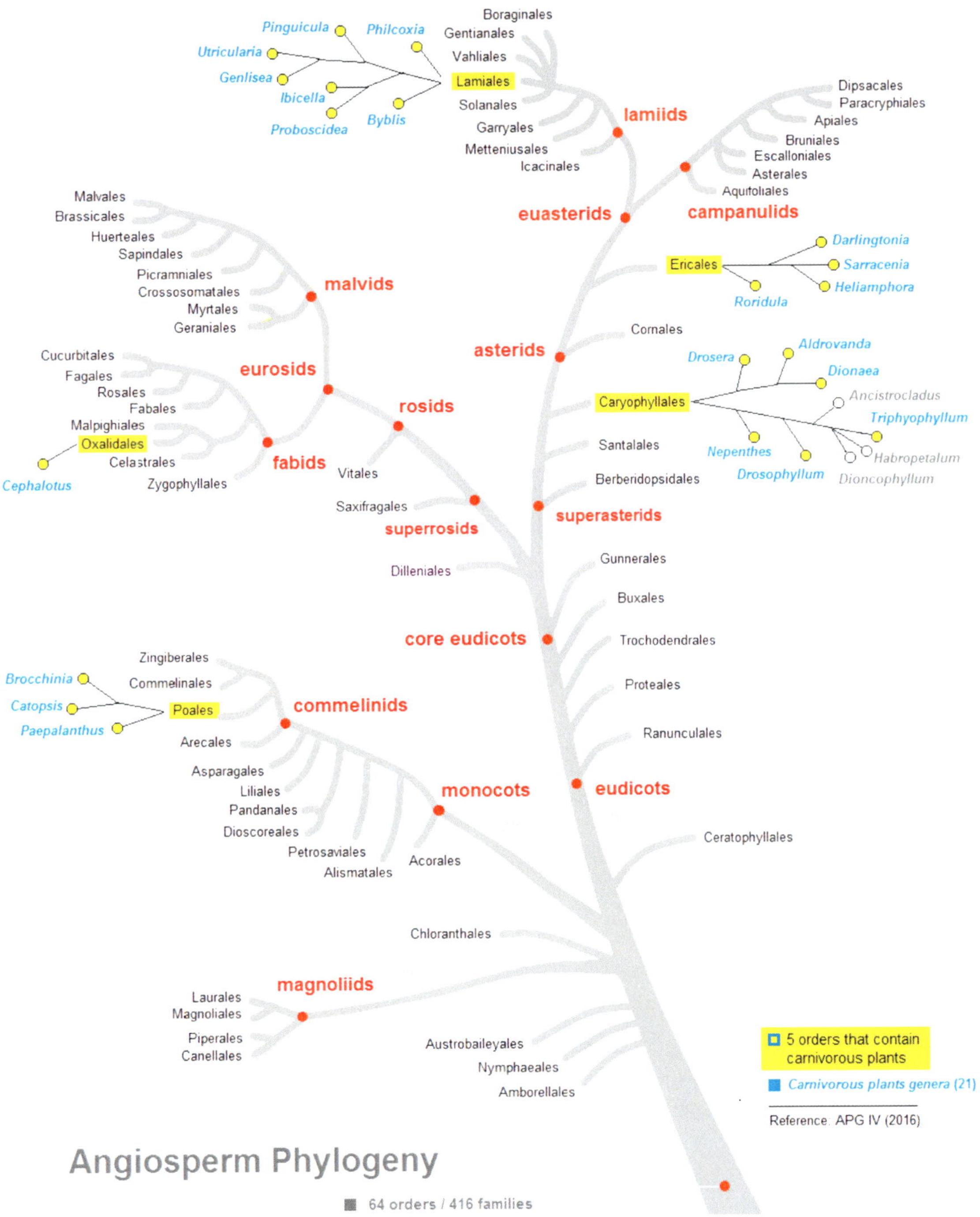

Boraginales
Gentianales
Vahliales
Lamiales
Solanales
Garryales
Metteniusales
Icacinales
Pinguicula
Philcoxia
Utricularia
Genlisea
Ibicella
Byblis
Proboscidea
lamiids
Dipsacales
Paracryphiales
Apiales
Bruniales
Escalloniales
Asterales
Aquifoliales
euasterids
campanulids
Malvales
Brassicales
Huerteales
Sapindales
Picramniales
Crossosomatales
Myrtales
Geraniales
malvids
Ericales
Darlingtonia
Sarracenia
Heliamphora
Roridula
Cornales
asterids
Drosera
Aldrovanda
Dionaea
Ancistrocladus
Triphyophyllum
Caryophyllales
Nepenthes
Habropetalum
Drosophyllum
Dioncophyllum
Santalales
Berberidopsidales
Cucurbitales
Fagales
Rosales
Fabales
Malpighiales
Oxalidales
Celastrales
Cephalotus
Zygophyllales
eurosids
rosids
fabids
Vitales
Saxifragales
superrosids
superasterids
Dilleniales
Gunnerales
Buxales
core eudicots
Trochodendrales
Proteales
Ranunculales
Zingiberales
Commelinales
Brocchinia
Catopsis
Poales
Paepalanthus
commelinids
Arecales
Asparagales
Liliales
Pandanales
Dioscoreales
Petrosaviales
Alismatales
Acorales
monocots
eudicots
Ceratophyllales
Chloranthales
magnoliids
Laurales
Magnoliales
Piperales
Canellales
Austrobaileyales
Nymphaeales
Amborellales
5 orders that contain carnivorous plants
Carnivorous plants genera (21)
Reference: APG IV (2016)
Angiosperm Phylogeny
64 orders / 416 families

group of plants, bladderworts, invented this trap, though the genus as a whole they achieved tremendous success with well over 200 species worldwide. *Utricularia* (and closely related *Genlisea*) belongs to the Lentibulariaceae family as does *Pinguicula*. The adhesive method of butterworts is not effective in the watery world, and the scheme had to be modified. The current conjecture is that the inwardly rolled-up version of *Pinguicula*'s adhesive leaves growing into the wet substrate (and replacing the true root) led to the present-day underwater traps of *Genlisea* and then eventually those of *Utricularia*.

Phylogenetic Tree

The tree depicting all 64 orders of angiosperms provides a bird's-eye view of the bewildering diversity of flowering plants and helps us place carnivorous plant occurrences in an evolutionary perspective. In the five orders that contain carnivorous plants, carnivorous genera (only) are expanded. One cannot escape the impression that, in the overall landscape of flowering plants on Earth, plant carnivory is a localized phenomenon. After all, only five orders contain carnivorous taxa. Poales is a group of monocots containing a few, very primitive carnivorous species. Oxalidales contains only one pitfall (*Cephalotus*). The rest occurs in the remaining three orders. Within these orders, carnivorous plants tend to cluster together. Even some pseudo-carnivores (à la *Ibicella*) occur fairly close to well-established carnivorous taxa. This may imply that some of these carnivores are in the process of evolving into or out of carnivory.

One observation to be made is that, in both Caryophyllales and Lamiales (in Lentibulariaceae), carnivory started out as an adhesive trap — and in time moved on to more complex, drastically different trapping schemes (snap-trap & suction trap). It is intriguing to note that, in both of these orders, the change appears to have coincided with the transition into the aquatic environment.

Pitcher Plants

GENUS *Sarracenia*
FAMILY **Sarraceniaceae**

There are eight species of eastern North American pitcher plants occurring in the Atlantic coastal plains of North America. Seven species are confined to

the eastern and southeastern part of the United States. The remaining one species extends its distribution all the way northward deep into a large part of southern eastern and central Canada.

The plants are typically found in marshy savannas, pine forests and marl fens, and along the stream and river bank for some isolated populations. The soil is constantly wet, generally acidic, and low in nutrients. In the south, the plants often grow with two or more species sharing the same habitat. This sympatry results in various natural hybridization.

In the early eighteenth century, a Canadian physiologist/naturalist Michel Sarrazin sent pitcher plant specimens to the renown French botanist Joseph Pitton de Tournefort. The genus name *Sarracenia* was adopted in honor of Dr. Sarrazin by Carolus Linnaeus (1737), who described the species based on Tournefort's original account. The common name for the genus came from their tubular leaves which retain some liquid.

A peculiar structure of the pitcher leaves, along with their striking appearance, has long attracted the attention of people. Linnaeus, the father of taxonomy, is one of the eighteenth-century botanists who believed — erroneously, as many others did at the time — that the lid of a pitcher is capable of movement in order to conserve the water within. Another botanist, Mark Catesby, thought that the pitchers were intended to provide a merciful refuge for poor insects fleeing from animal predators. It was not until the beginning of the nineteenth century that more serious observations started to reveal the true carnivorous nature of the plants.

TOP: *Sarracenia leucophylla* in flower, late April, in Florida.
OPPOSITE: *S. leucophylla*, in late April, southern Alabama.

DESCRIPTION

Pitcher plants are herbaceous perennials consisting of a rhizome with thick fibrous roots. The hollow trap leaves arise directly from the rhizome. The pitcher leaves form a rosette and are erect in most species but are decumbent in some. A lid develops at the upper end of the pitcher. The lid is typically reflected over the pitcher opening, but may grow to form a hood in some species. The mature pitchers range in height from several centimeters to a meter, depending on the species and growing conditions.

> ❝ **In *Sarracenia*, a pitcher is formed by folding a flat leaf inward and fusing the edges together.**

The pitcher leaf of *Sarracenia* is epiascidiate in ontogeny: A pitcher is formed by folding the flat leaf inward and fusing the edges together. The inner surface of the pitcher tube corresponds to the adaxial (upper) surface in an epiascidiate leaf.

Phyllodia The plants produce phyllodia. These are a flat, sword-like leaf made of a prominent ala, without a hollow tube. Phyllodia are generally produced as a winter resting leaf and, in some species, during the summer semi-dormancy. Some species produce phyllodia predictably and consistently while others produce them irregularly in response to some stressful conditions to enhance photosynthesis.

ABOVE: A colony of *Sarracenia alata* in a pine savanna. In early July, southern Mississippi. OPPOSITE: *Sarracenia purpurea* ssp. *purpurea* in a fen in northern Michigan. Flowering ended a month earlier and seeds maturing, mid-July.

" Pitcher plant flowers are designed to encourage cross-pollination.

The flower of *Sarracenia flava*, with two petals removed. Note two of the five stigmatic lobes visible on the inner side of the cutaway umbrella style.

The basic flower structure of pitcher plants is the same among all species, having five petals, five sepals and three bracts. In early spring, a tall scape emerging from the rosette center supports a solitary, nodding flower with a showy coloration and rather odd appearance. The actinomorphic (radially symmetric) flower has a unique morphology that suggests an advanced floral adaptation for pollinator interaction.

Nature often provides various mechanisms that prevent a flower from being fertilized by its own pollen. In pitcher plants, the flowers are protogynous, with flower's stigma becoming receptive prior to pollen shedding. Also, the obvious exposure of pollen-receiving stigma outside the closed corolla chamber appears to encourage cross-pollination.

When a flower opens in a pendulous position, the umbrella-shaped style hangs upside down, with five umbrella points curving upward. At the tip of each point grows a tiny stigma lobe projecting inward. Many stamens surround the base of a round, five-chambered ovary. Five petals grow downward and press against the flattened bottom of the umbrella style, forming a sealed corolla chamber encasing the stamens within. The petal then sharply bends outward and rolls out through one of the five cut-away arcs of the umbrella, leaving the remaining lower two-thirds of the petal dangling outside the umbrel-la style. Each petal is pressed tightly against an umbrella arc such that a small notch is formed on each side of a petal. This is noticeable if a petal is removed from the flower. These small indentations are the only visible clue for a pollinator seeking entry into the flower.

This arrangement structurally separates pollen-producing anthers from the stigmas located outside the corolla chamber. Suppose an insect pollinator lands on a flower. When it tries to find an entrance to the corolla in search of nectar, a stigma at one of the umbrella points — located just above the parting of two neighboring petals — is bound to be brushed and the pollen from the previously visited flowers deposited. Once inside the corolla chamber, the insect seeks nectar at the

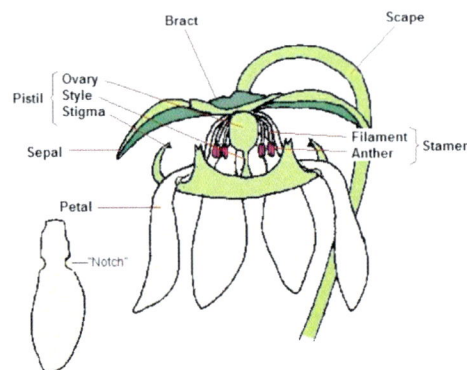

Pitcher Plant Flower

TOP: The red flowers of *Sarracenia leucophylla* in a coastal savanna along the Gulf Coast. Note that many different pitcher plant species are sharing the open, coastal habitat. In late April, in the Florida Panhandle. ABOVE: A hybrid flower. The small stigmatic lobes are seen projecting inward from the tips of the umbrella style.

base of the stamens. As it does so, the insect collects ample amount of pollen which probably has been accumulated also on the umbrella floor. When the insect is ready to leave the flower, it is likely to push one of the hanging petals from the arc of the umbrella, rather than retrace the same petal parting near the stigma. This way, the pollinator, now with the flower's own pollen, does not touch the stigma again, thus avoiding self-pollination.

Seeds mature in July through September in the southeastern United States habitats. The ovary splits and the seeds are dispersed. The *Sarracenia* seeds are fairly large measuring 2 mm or so. The seeds generally germinate right after dispersal without requiring stratification, but in cultivation, seeds may be exposed to a period of cold temperature in a damp medium to enhance germination.

Following twin cotyledons, a seedling produces a tiny juvenile leaf with a tube. The juvenile leaf looks more or less the same for all species and does not exhibit distinct leaf characteristics of the parent for a year or two. Then, quite spontaneously, the plant produces an adult pitcher leaf. Pitcher plants usually mature from seedling to flowering age in four to five years. The plants also propagate asexually by crown division and underground rhizome.

TRAP STRUCTURE AND ATTRACTION

Although a specific pitcher shape is characteristic for each species, the basic trap structure remains the same throughout the genus. The hollow leaves of *Sarracenia* are carefully constructed pitfalls designed to attract and capture small animal prey. The colorful leaves are often mistaken for flowers by visiting insects and uninitiated human observers alike. In fact, pitcher leaves have evolved to exhibit alluring elements of real flowers — striking colors, patterns and copious nectar secretions. Some pitcher plants release a fragrance in addition to nectar production. The ultraviolet photography of the pitcher reveals distinct UV absorption patterns as are commonly seen in many an insect-pollinated flower. Much of the outer surface of the pitcher is scattered with nectar glands forming a nectar trail. This attracts crawling creatures on the ground and guides them to the pitcher opening.

The pitcher lid may not prevent rainwater from entering the pitcher due to a varying degree of reflection in some species. The lid does provide a convenient landing site and feeding ground for flying insects. Trying to lick nectar, however, is a risky business for venturing insects, as we shall see. Studies have shown that some pitcher plants produce nectar laced with a neurotoxin (coniine) that intoxicates prey.

In *Sarracenia*, many species show broad color variation. Distinct color variants that exist in the wild population are a reflection that, in the adaptation to carnivory, the selective pressure had a strong influence on the visual appearance of the arthropod-trapping foliage in the genus.

Zones In view of prey trapping and the subsequent absorption of nutrients, the inner surface of the pitcher is divided into several zones. The lid is Zone 1, the attracting zone, characterized by vibrant colors and numerous nectaries (nectar glands). The undersurface of the lid is lined with stiff, short hairs all pointing in the direction of the pitcher opening. The foothold is precarious and the hairs encourage an insect to move closer toward the pitcher opening. The slightest slippage can spell disaster. This is also where the ultraviolet absorption patterns are most pronounced.

Zone 2 is the conduction zone. This includes the rolled edge of the pitcher mouth and the inner wall that extends for some distance below. Nectar secretions are most abundant in

the rolled margin ("nectar roll") and nectar drops often form on the edge of the peristome. The inner wall abruptly changes to a surface of overlapping trichomes each narrowing down to a point. This makes the wall surface slick. This is where the insect is likely to lose its footing and plummet into the trap.

Next comes Zone 3, the glandular zone. Many glands are found embedded in the inner wall. The wall surface provides no foothold for an unfortunate insect attempting to escape. This zone continues half way down toward the pitcher bottom.

In Zone 4, the retention zone, the inner wall is lined with long, downward-pointing hairs. This hinders ascent and the

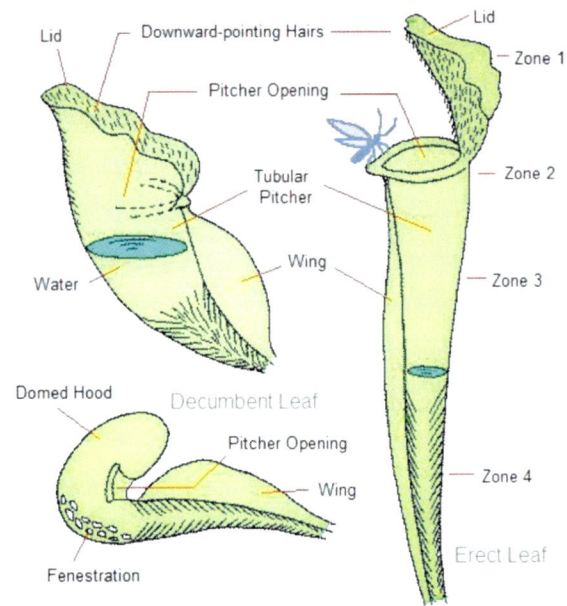

Pitcher Plant Leaves

prey is retained in this region. The wall has numerous glands. Typically, the pitcher liquid covers this area and digestion as well as absorption of dissolved nutrients takes place here.

Wind Damage Generally, in erect species, only the lower quarter of the pitcher is filled with water. Tall species, like *Sarracenia flava,* are particularly susceptible to the forces of wind if the pitcher gets top-heavy due to excess water after rain. Strong winds can easily topple the slender pitcher in the wild. Once bent, the pitchers are unlikely to recover from the physical damage. Ironically, ferocious storms and hurricanes are commonplace in the coastal regions of the southeastern United States where these plants abound.

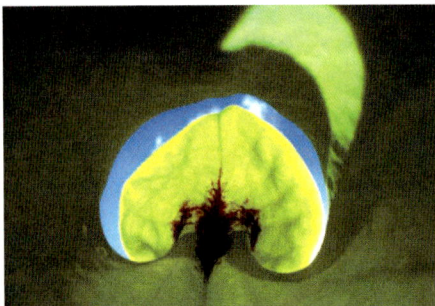

Sarracenia flava.

Weighing the benefit of abundant nectar over a deadly plunge, a fly exhibits unwavering faith in the power of the adhesive pads on its feet — in the middle of "Zone 2" designed to conduct prey to the trap. *Sarracenia oreophila* in northern Alabama, early May.

TOP: A drowned moth creating a mess in the pitcher liquid of *Sarracenia flava*. In early July, Florida. MIDDLE & BOTTOM: A spider trapped in the pitcher of *S. flava*, in early July, Florida.

DIGESTION

Toward the end of the leaf development, digestive glands secrete a small amount of liquid into the pitcher cavity while the pitcher is still closed. This sterile liquid is slightly acidic. When the pitcher has matured and the lid opens, the trap is ready. In this passive pitfall trap, digestion of prey takes place in the solution at the pitcher bottom. The prey literally jumps into the pool of pre-formulated digestive fluid. While rainwater may dilute the pitcher liquid, the acidity is known to be maintained, at least in a younger leaf. Studies have shown that chemical stimulation by beef broth results in a multiple increase of fluid secretions in an unopened pitcher.

The surface tension of the pitcher liquid is measured to be lower than that of water. This promotes swift drowning of insect prey, acting as a wetting agent on the otherwise water-repelling insect body.

Enzymes Researchers have been trying to determine the origin of enzymes present during the digestive process. Studies have confirmed protease secretions in an unopened pitcher in some species, though it is generally believed that the digestion is heavily assisted by bacteria externally introduced with the prey during much of the pitcher leaf's lifecycle. Other enzymes detected in the pitcher liquid include esterase, acid phosphatase and amylase. The products of digestion are promptly absorbed by the digestive glands on the inner leaf surface.

ABOVE: A *Sarracenia flava* colony in a coastal savanna. In Florida, early July. PAGE SPREAD: A bug's-eye view of a *S. flava* pitcher, looking up from the bottom of a pitcher.

Inquilines In a complex web of the ecosystem, the pitcher plant liquid is inhabited by a multitude of creatures who spend all or part of their life in the pitcher. These commensal organisms are not harmed by digestive enzymes present in the liquid.

In a northern habitat of *Sarracenia purpurea*, researchers have identified many such organisms. These inquilines include mites (*Anoetus gibsoni* and others) and mosquito larvae of three *Diptera* species, as well as rotifers, nematodes, copepods and various protozoa. Studies have also found

" Inquilines are an integral part of the whole digestive scheme in *Sarracenia*.

that these *Diptera* larvae participate in the overall digestion sequence by assuming different task divisions. The larvae of *Blaesoxipha fletcheri* first attack newly captured prey floating on the surface. The decomposed victims are then consumed by free-swimming *Wyeomyia smithii*. The remains accumulated on the pitcher bottom are consumed by *Metriocnemus knabi*. The larvae feed on protozoa and bacteria also. An adult mosquito of *Wyeomyia smithii* is known to deposit eggs only in pitcher plants.

Mutualism Observations in the field show that many visitors to the pitchers — presumably having enjoyed nectar offerings — do leave the plants without being caught, and digested. This leads to a fascinating ecological model in which pitcher plants provide a vital service for certain insect communities in a mutually beneficial manner, thus forming a "mutualism": Insect communities benefit from nectar provided by pitcher plants — in exchange for a small portion of the communities being sacrificed as prey.

Pests / Infestation Even *insectivorous plants* are not entirely immune against insect attacks. Adult *Exyra* moths are known to deposit eggs in the upper portion of a pitcher. As the eggs hatch, larvae of *Exyra* moth emerge. The top of the pitcher is often bent as the caterpillar of this moth eats the pitcher inside out with voracious appetite. A bent-over pitcher leaf is an unmistakable sign of *Exyra* infestation. In the safety of sealed pitcher enclosure, the larvae pupate. In time, a new moth is born and it flies away if only to a nearby, fresh pitcher and the cycle repeats. The *Exyra* species associated with pitcher plants include *E. ridingsii* (in *S. purpurea* and *S. flava*), *E. rolandiana* (in *S. purpurea*) and *E. semicrocea*.

A sighting of the green lynx spider is common on *Sarracenia* pitchers. The lynx spiders have a worldwide distribution. The green lynx spider (*Peucetia viridans*) is one of the largest in the family and occurs widely in North America, often found on leaves and flowers. A nimble and aggressive hunter, the spider finds pitchers of *Sarracenia* a convenient hunting ground. Hiding near the pitcher mouth, the keen-eyed spider ambushes its prey. Sometimes a trapped insect is stolen from the pitcher. In the act of robbery, the spider does not seem to fall victim to the slippery pitcher trap.

LEFT: A green lynx spider on the lid of a *Sarracenia flava* pitcher. In July, North Carolina. RIGHT: A tiny spider on the inner lid surface of a *S. leucophylla* pitcher. In southern Alabama, early May.

Mutualism: *Sarracenia leucophylla* and a wasp. "Hope you enjoyed the meal — the *payment* is due by the end of summer." In early May, southern Alabama.

Sarracenia alata

The plants are found in two separate ranges: an area around coastal Mississippi just extending into neighboring Alabama and Louisiana, and an inland area covering western Louisiana and the adjacent eastern Texas.

The pitcher of *Sarracenia alata*, that grows to 75 cm, represents a typical morphology of an erect species, having a tall, tubular leaf with a reflecting lid at the peristome. The pitcher is quite similar in appearance to that of *S. rubra,* causing misidentification in the field. Both *S. rubra* and, to a lesser extent, *S. alata* are variable in their pitcher characters. *S. alata* pitchers are generally light green with some venations but various color variants are noted in the field, and five varieties have been formally described: variety *atrorubra*, variety *cuprea*, variety *nigropurpurea*, variety *ornata*, and variety *rubrioperculata*. The color of the pitcher tends to grow darker and reddish as the pitcher ages. *S. alata* may produce phyllodia in mid-summer, and then in autumn for winter dormancy, though not always consistently.

Flowering starts in late March and the pale-yellow flowers of *S. alata* bloom profusely into April. New leaves emerge well after anthesis, providing a clear separation between pollination and prey trapping. The flower scape stands 30-40 cm tall and new pitchers easily surpass the flowers in height.

ABOVE: Heavily veined, new pitchers of *Sarracenia alata*. In early May, southern Mississippi. RIGHT: Typical, greenish pitchers of *S. alata*. In early May, southern Mississippi.

ABOVE: A colony of *Sarracenia alata* in a Gulf Coastal savanna. In early July, Mississippi.

LEFT: Deep-red *Sarracenia alata* pitchers in summer. In July, in Mississippi. FAR-LEFT TOP: Right after flowering, in early May, Mississippi. FAR-LEFT BOTTOM: The Autumn color of maturing fruit capsules. In July, Mississippi.

55

Pale-yellow petals of *Sarracenia alata* swinging in a spring breeze. Ample pollen deposits accumulate on the floor of the inverted umbrella-shape style for a pollinator to bathe in. In late April, southern Mississippi.

TOP: Sprouting flower buds of *Sarracenia alata*, in early March, southern Mississippi. BOTTOM: A field of *S. alata* right after flowering, with new pitcher leaves just emerging. In late April, southern Mississippi.

RIGHT: A mature seed capsule of *Sarracenia alata*. The pod of all species splits from the frontal tip of the capsule, with the exception of *S. leucophylla* whose pod splits from the back.

Sarracenia rubra

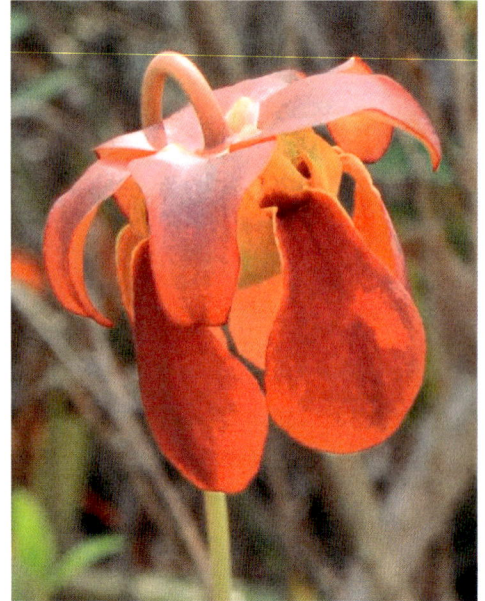

The distribution of *Sarracenia rubra* consists of several areas in the southeastern United States. The pitchers of *S. rubra* are highly variable among different populations, and there are differing opinions regarding its classification. Here, following the taxonomic treatment of D. Schnell (2002), the entire *S. rubra* complex is described under one species, with five subspecies. All these infraspecific variants maintain disjunct populations and are geographically separated. With the exception of subspecies *rubra* that has a relatively wide distribution, each subspecies only occurs in a restricted geographic range.

S. rubra in general is a relatively small plant with a typical pitcher leaf 15-45 cm tall. The pitcher exterior for all subspecies shows fine pubescence (1 mm hairs). All variants share a red flower color and almost identical flower morphology.

Sarracenia rubra **subspecies** *rubra* Known as the sweet pitcher plant, this taxon occurs in the original species type distribution along the Atlantic coastal plain of North Carolina, South Carolina and Georgia.

Sarracenia rubra **subspecies** *gulfensis* The populations are found in western Florida. The plant often prefers wet edges of creeks and lakes, and are not likely to be found in the coastal savanna where many *Sarracenia* species grow sympatrically. The flowers bloom in early April. A giant form of this subspecies produces pitchers reaching 70 cm tall.

Sarracenia rubra **subspecies** *wherryi* Occurring in southwestern Alabama, the plants grow in pine savannas, sometimes sparsely intermingled with *S. leucophylla* and *S. psittacina*, resulting in natural hybridization. Flowers in early April.

Sarracenia rubra subspecies gulfensis

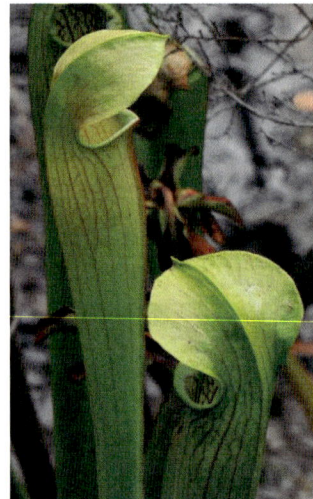

TOP: The red flower of *Sarracenia rubra* ssp. *gulfensis*. In early May, Florida.
LEFT & ABOVE: *S. rubra* ssp. *gulfensis*, a "giant" form, growing near the edge of a pond, with pitchers reaching 70 cm tall. Early May, in the western Florida Panhandle.

A typical pitcher of *Sarracenia rubra* ssp. *gulfensis*
(30 cm tall). Note a velvet-like pubescence on the
pitcher exterior, that is common to all *rubra*
complex. In early May, in the western Florida
Panhandle. INSERT: A pink-flowered orchid,
Pogonia ophioglossoides, growing abundantly
alongside *S. rubra* ssp. *gulfensis*.

A well-veined pitcher of *Sarracenia rubra* ssp. *wherryi*. In early May, southern Alabama. INSERT: Blooming *Cleistes bifaria*, an orchid native to the southeastern part of the U.S., often seen in pitcher plant habitats.

A clump of *Sarracenia rubra* ssp. *wherryi*, with a prominent cupper-tinted pitcher lid exterior. The pitchers grow to 40 cm tall. In early May, southern Alabama.

Sarracenia rubra subspecies *wherryi*

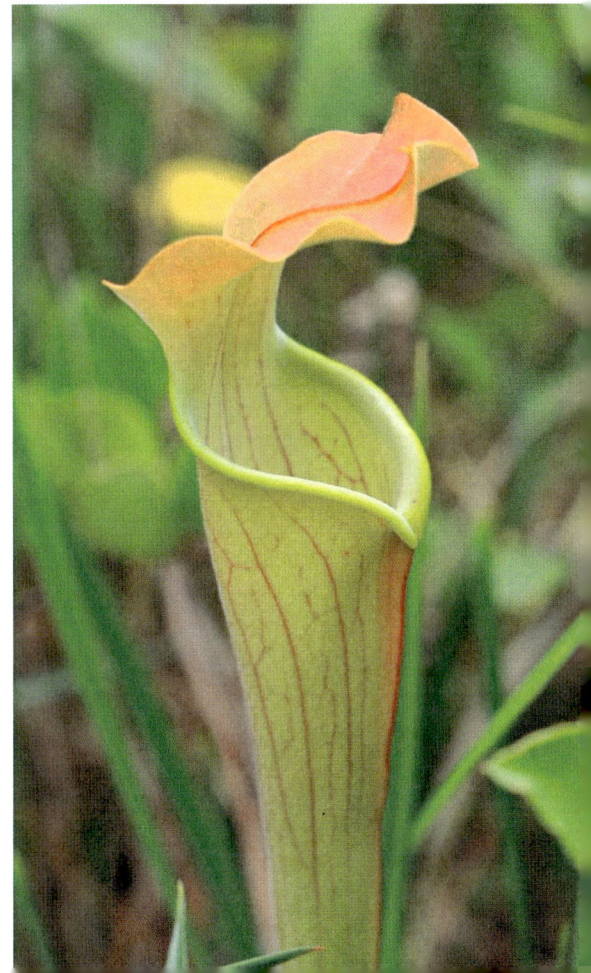

LEFT: Slender pitchers of *Sarracenia rubra* ssp. *wherryi*. In early May, southern Alabama.
RIGHT: The pitcher of *S. rubra* ssp. *wherryi*. Note the lid with wavy margins and prominent cupper-tinted exterior. In early May, southern Alabama.

Sarracenia rubra subspecies *alabamensis* Locally known as the canebrake pitcher plant, the populations are confined to several sites scattered over three counties in central Alabama. No other pitcher plants grow in the region. Flowering peaks in mid-to-late April. Subspecies *alabamensis* is placed on the federal endangered species list.

Sarracenia rubra subspecies *jonesii* Only found in a small area along the North Carolina and South Carolina border, the pitcher shape very much resembles that of *S. alata*. Subspecies *jonesii* is on the federal endangered list.

Sarracenia rubra subspecies *alabamensis*

LEFT: Alabama canebrake pitcher plants in flower, with the flower scapes growing to 45 cm tall. In early May, central Alabama. *Sarracenia rubra* ssp. *alabamensis* is on the federal endangered species list. BELOW: A colony of *S. rubra* ssp. *alabamensis*, showing stressed growth due to low precipitation of the previous year, with spring pitchers barely reaching 15 cm. In early May, central Alabama.

Vigorous growth of *Sarracenia rubra* ssp. *alabamensis*, with pitchers reaching respectable 35 cm in length. The pitcher lid exterior of subspecies *alabamensis* also assumes a copper-tinted coloration, though slightly weaker than that of subspecies *wherryi*. The moist forest floor is covered with tiny white blossoms of dwarf huckleberry (*Gaylussacia dumosa*), a low-lying shrub belonging to the heath family. In early May, central Alabama.

Sarracenia rubra subspecies alabamensis

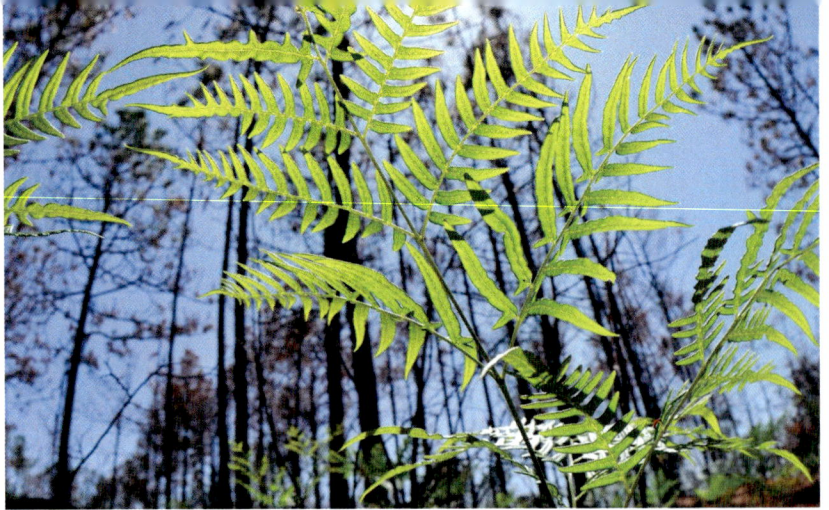

ABOVE: A habitat of *Sarracenia rubra* ssp. *alabamensis* in central Alabama, early May. The forest harbors various types of ferns, adding to the rich vegetation of the mountain. LEFT: A new spring pitcher of *S. rubra* ssp. *alabamensis*. In early May, central Alabama. BELOW: A fly enjoys a fresh mountain air atop a newly opened spring leaf of *S. rubra* ssp. *alabamensis*. In early May, central Alabama.

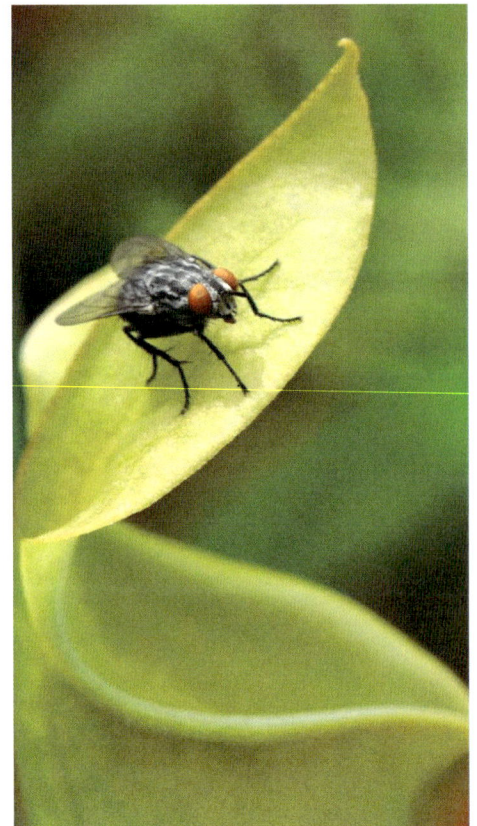

Blossoming red flowers of *Sarracenia rubra* ssp. *alabamensis* immediately after the passage of torrential shower. Flowering generally peaks around mid-to-late April in this subspecies. In early May, in central Alabama.

Sarracenia oreophila

Found in a few disjunct locations in northern Alabama and the Georgia-North Carolina border, the species is in danger of extinction.

Found in a handful of disjunct locations scattered in northern Alabama and the Georgia-North Carolina border, the species is in danger of extinction and is federally protected. Commonly known as the green pitcher plant, the pitcher of *Sarracenia oreophila* bears a close resemblance to that of *S. flava*. In fact, the plant was considered a variant of *S. flava* until 1933 when E. T. Wherry formally described it as a separate species. Compared with *S. flava*, the pitcher of *S. oreophila* has a wider neck column and its flower has a less pubescent umbrella style. In spite of its apparent similarity to *S. flava*, a recent DNA comparison suggests that *S. oreophila* is more closely allied to *S. alata* and *S. rubra*, while *S. flava*, *S. minor* and *S. psittacina* are equally closely allied.

The pitchers are generally light green, with varying degree of red to brownish venation on the inner tube surface. Some pitchers may assume a light cupper tint over the lid exterior. One color variant, *S. oreophila* variety *ornata*, is formally recognized that exhibits deep-red pitcher exterior venation. As in other *Sarracenia* species, the pitcher darkens in color as it ages. Slightly shorter than *S. flava*, the pitcher grows to 70 cm tall. The plant tends to form a clump by underground rhizomes. In Alabama, there seems to be no other carnivorous plants growing with *S. oreophila*.

The yellowish-green flowers start to bloom in early May in northern Alabama. New spring pitchers are fully grown and functional at the time of flowering. The temporal separation between flowers and pitchers designed to provide a margin of safety for pollinators is cheerfully violated in *S. oreophila*. The montane habitat experiences short spring before hot, dry summer forces the plant into semi-dormancy. Pressed for time, sex and dinner proceed concurrently in this species.

As the condition ameliorates in late summer, the plant resumes its growth with the second set of pitchers somewhat smaller than the spring growth. The plant produces characteristic, short, curved phyllodia in autumn in preparation for long winter dormancy.

TOP: The greenish-yellow flower of *Sarracenia oreophila* with five petals, five sepals and three bracts. Early May in northern Alabama. BOTTOM: The production of short, sharply curved phyllodia is a characteristic feature of this species. In this picture taken in May, the spring flooding of the river submerged the plants, causing the deterred development of new spring leaves this year. In northern Alabama. OPPOSITE: *S. oreophila* maintains a modest colony in this montane habitat, surrounded by giant ferns and encroaching bushes. With the scattered growth of some *Sphagnum* moss nearby, there are no other carnivorous plants found in this site. In early May, northern Alabama.

" **Pollinator safety is of little concern in the overall scheme of things in *Sarracenia oreophila*.**

ABOVE: A rush of new pitchers of *Sarracenia oreophila* covering the colony at the time of flowering — no *temporal* separation. Fully functional pitchers surround the yellow flowers which, incidentally, stand more or less at the same height as the pitcher openings — no *spatial* separation. In this species, pollinators visit the flower at their own risk. Furthermore, the yellowish flower fails to create a discernable visual contrast against the backdrop of light-green pitcher leaves. In early May, northern Alabama. RIGHT TOP: A pitcher with veins. RIGHT BOTTOM: A trapped bunch in the pitcher. In early May, northern Alabama.

Sarracenia leucophylla

Sarracenia leucophylla grows on the Gulf Coast from southwestern Georgia through the Florida Panhandle, Alabama and just into Mississippi.

The erect pitchers of *S. leucophylla* grow to one meter in height. The pitchers become white toward the top with red or green to greenish-brown venation against the white background. In the field, broad color variations are quite noticeable, with varying color intensity. Some specimens exhibit truly deep-reddish coloration while, at the other extreme, almost entirely white pitchers are also seen. A white variant with no discernable venation on the interior of the pitcher opening is formally described as *S. leucophylla* variety *alba.*

Flowering starts in late March and continues into April. The red flowers brighten the spring savanna visible from distance. The petals last for a few weeks and new leaves emerge soon after. In late April, the new pitcher leaves may be seen alongside still attractive red flowers.

Sometimes the plant produces phyllodia in summer. These are flat leaves totally lacking the pitcher tube. Then in late summer, the second crop of pitchers are produced, often larger and more robust than the spring leaves. Occasional winter phyllodia may be produced.

The seed pods of *S. leucophylla* split from the rear rather than front. This is the only species to do so in the genus. *S. leucophylla* is arguably the most attractive of all pitcher plants, and the leaves are in high demand in cut-flower trade.

ABOVE: A new spring pitcher of *Sarracenia leucophylla.* The conspicuous network of red veins over the white pitcher top creates an effective visual lure for visiting insects. Note that the underside of the lid is covered with fine, downward-pointing hairs for prey capture. In early May, southern Alabama. RIGHT: Red-heavy pitchers growing among the more typical, lighter-red population. In early May, southern Alabama.

Bright-red flowers of *Sarracenia leucophylla* blooming in a coastal savanna along the Gulf Coast. New pitchers are just emerging. Open coastal savannas like this one are increasingly rare. In Florida, late April.

Pitcher color variations of *Sarracenia leucophylla*, from extremely-deep-red to green to white. The color comparison ought to be done with pitchers of the same age — preferably with fresh, new pitchers — since the color tends to darken, sometimes drastically, as the pitcher ages. In early May, southern Alabama.

71

"The color of deception.

A breath-taking spectacle of a dense stand of *Sarracenia leucophylla* formed in an open savanna sparsely populated by longleaf pines (*Pinus palustris*) in southern Alabama. The colony explodes with a colorful exposé of pitcher leaves in early May.

The main function of a leaf is photosynthesis and leaves normally assume a green color for this purpose. In carnivorous plants, a leaf takes on the additional task of trapping prey for consumption. With it comes a need for attracting insects to the leaf.

It is no coincidence that the leaves of carnivorous plants mimic flowers in terms of their beauty — exploiting all alluring elements of real flowers. This is nowhere more vividly expressed than in the field covered with thousands of tall, colorful pitchers of *S. leucophylla*.

Sarracenia flava

TOP: *Sarracenia flava* var. *rubricorpora* intermingled in a more common variety *rugelii* population in the Florida Panhandle, early May. BOTTOM: *S. flava* var. *flava* found growing in the Florida Panhandle, early May. This is a dominant variety in the northern range of the *S. flava* distribution, but less common in Florida. OPPOSITE: A colony of *S. flava* var. *rugelii* in the Florida Panhandle, in early May. The pitchers often reach one meter in height.

The distribution of *Sarracenia flava* extends along the Atlantic coast from the southeastern corner of Virginia all the way south to southern Alabama in the Gulf Coast.

S. flava is a tall plant with its erect pitchers sometimes exceeding one meter. The common name, yellow trumpet, describes its slim, handsome-looking pitcher shape. The plants are typically found in a grassy savanna often forming a large colony. The plant produces winter phyllodia which are flat and lack pitcher tubes.

Flowering occurs early in mid-March in its southern range (in the Florida Panhandle) and in May in the northern localities (around North Carolina). The bright, deep-yellow flowers are one of the largest and most striking in the genus. In this species the old pitchers die back completely during winter months, and in spring, the flower scapes pierce through the piles of dead leaves lying on the ground. The forest of bright yellow blossoms covers the colony, creating a spectacular floral display in the spring wetland. Immediately after flowering, a rush of new leaves grow well passing the flower scapes. The tall spring pitchers ready themselves as the flower petals fall off after fertilization.

A wide range of pitcher color variation exists in this species and several infraspecific variants are formally described at variety level.

Sarracenia flava **variety *flava*** A deep-red blotch on the throat (inner column of the lid) with radiating venation around it characterizes this variant. Common in the Atlantic coastal plain.

Sarracenia flava **variety *maxima*** The pitcher is totally green with no red venation. Found in North Carolina but uncommon elsewhere.

Sarracenia flava **variety *cuprea*** Characterized by the deep-cupper external lid surface, this is common in the northern range.

Sarracenia flava **variety *atropurpurea*** The pitcher is almost totally red. This is not a common variety.

Sarracenia flava **variety *ornata*** Having heavy venation over the pitcher (including the lid), this is common in the northern range but less common in Florida.

Sarracenia flava **variety *rubricorpora*** Distinguished by a deep-red external pitcher tube, this conspicuous variety is confined to the Florida Panhandle.

Sarracenia flava **variety *rugelii*** With a deep-red blotch on the throat but without emanating venation, this is a predominant plant in Florida but rare in the north.

Bright yellow blossoms of *Sarracenia flava* var. *maxima* plants covering a bushy field in coastal North Carolina, late April. The large, deep-yellow flowers are one of the largest in the genus, measuring 10 cm across. In this species, the old leaves typically die out completely in winter. Come spring, the flower scapes pierce through the pile of dead leaves, bearing a nodding flower at the tip.

> **Bright yellow blossoms in a coastal North Carolina bog.**

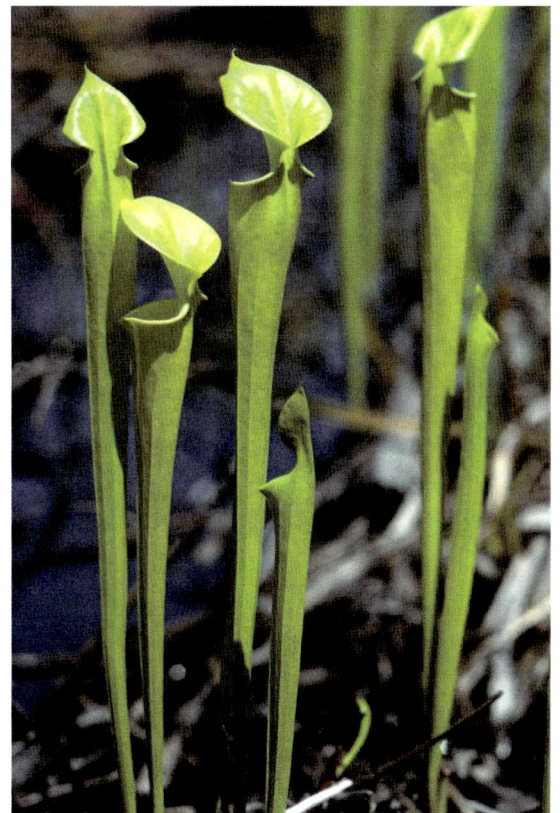

RIGHT TOP: As the flowers reach anthesis, new leaves hurriedly emerge from the rosette center, in preparation for the summer feast. In *Sarracenia flava*, the leaves easily surpass relatively short flower scapes. In late April, North Carolina. RIGHT BOTTOM: All-green pitchers of *S. flava* var. *maxima,* in early May, North Carolina.

RIGHT: A dense clump of *Sarracenia flava* var. *rugelii*, in early May, in Florida. A clump formation like this, sometimes reaching one meter in diameter, is quite typical in variety *rugelii* but is fairly rare in some other *S. flava* varieties. BELOW: *S. flava* var. *rugelii* growing in a grass-covered open field in a pine savanna. In early May, the Florida Panhandle.

ABOVE: *Sarracenia flava* var. *rubricorpora* — a lighter version. In early May, in Florida.

RIGHT: *Sarracenia flava* var. *ornata* (facing the camera), in early May, Florida. This is a *flava* variety with heavy venation over the entire pitcher. This variant is more common in the north.

Sarracenia minor

The *Sarracenia minor* distribution extends along the Atlantic coast, from the southeastern corner of North Carolina down south half way into the Florida peninsula.

The plant is relatively small with a typical pitcher size being less than 25 cm. However, a giant form is known from the Okefenokee Swamp in Georgia (*S. minor* variety *okefenokeensis*) that produces a pitcher reaching 100 cm tall. The plant generally prefers moist but somewhat drier soil in pine savannas, whereas the giant form is often found in much wetter conditions.

Commonly known as the hooded pitcher plant, a well-developed lid of *S. minor* overhangs the pitcher opening, forming a hood. Seen in a close-up, the hooded top of the pitcher evokes the feeling of a cartoon character, one of comical disposition.

The leaves are predominantly light green, assuming a varying degree of reddish tint toward the top. The undersurface of the lid is heavily lined with red venation, making the lid interior appear bright red. The back side of the upper pitcher is scattered with white areoles that help lighten the pitcher interior. For a winged insect alighting on the pitcher mouth, these white patches shine brightly, giving a false impression of exit to outside. In its attempt to fly through, the insect slams against the pitcher wall and tumbles into the depths of the pitcher bottom.

The greenish-yellow flowers bloom in early April into May. Spring leaves appear slightly before the flowers and the new pitchers are fully functional and attractive during flowering. There appears no consideration for pollinator safety in this species. The flower scape is often shorter than a pitcher. If a flying insect, carrying pollen from one flower to another, looks up, as it exits the flower, what it sees is brilliantly-lit red patches (lid undersides) against the blue sky, seductively positioned just above. Could it be that, in *S. minor,* the grand scheme of things is to have a cake and eat it too?

It is suggested that the primary diet of *S. minor* is ants, thus achieving prey-pollinator partitioning in spite of simultaneous production of flowers and traps. While ants may account for a large portion of its dietary supplement, field observations show frequent and enthusiastic visits of winged insects to the pitcher as well.

TOP: Flowering *Sarracenia minor* plants in Georgia, late April. The pale-yellow flowers bloom among already functional, new spring pitchers. The flower scapes are slightly shorter than the pitchers.
BOTTOM: The carefully contrived design of a *Sarracenia minor* pitcher. The bright red lid interior, visible from below, provides an effective lure. Strategically placed white patches on the back create illusion of an open gateway.

An insect's-eye view of the hooded pitcher interior, looking up from the bottom of the pitcher. The red-tinted ceiling of the hood shines brightly, as the white areoles scattered around the pitcher back provide illumination.

" A small chat on the grass.

ABOVE: Time for a small chat — *Sarracenia minor* growing on the grassy field near the Okefenokee Swamp in Georgia, early May. The plant typically prefers moist but somewhat dry area. Note a deep, reddish coloration on the pitcher exterior in this population. RIGHT: A wasp showing interest in the nectar offering. Considering the strategic placement of white areoles on the pitcher back, *S. minor*'s appetite for winged insects cannot be ruled out readily. In early May, Georgia. BELOW: *S. minor* in bloom in May, Georgia.

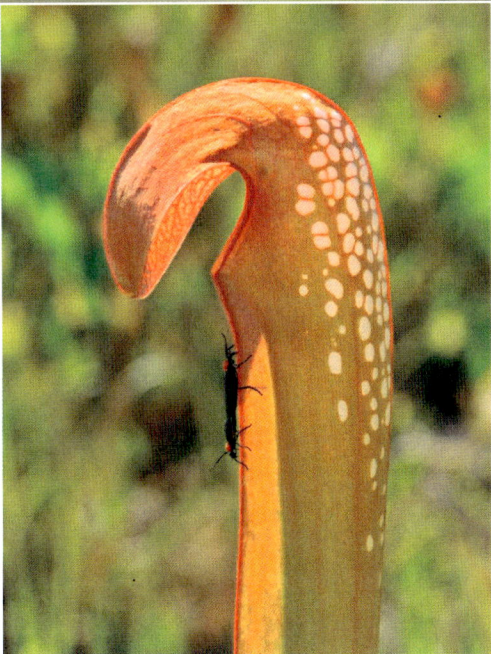

LEFT: A mating lovebug pair. The nectar trail along the ala (wing) of the pitcher brings crawling insects to the pitcher orifice where nectar secretions are the heaviest. ABOVE: The moment of a "joint" fall. The hard, slippery ridges of the pitcher orifice quite effectively conduct the prey into the depth of the pitfall trap. Another visitor is seen following the trail as the preceding couple undergoes a decisive plunge. In Georgia, early May.

Sarracenia psittacina

The pitcher mouth of a new *Sarracenia psittacina* leaf. White windows on the back help light up the pitcher interior. In early May, southern Alabama.

Sarracenia psittacina grows along the Gulf Coast, from southern Georgia westward just into western Louisiana. The plant seems to prefer a wetter habitat compared with other species.

The well-developed hood forms a dome over the pitcher opening, in much the same way as in the California pitcher plant (*Darlingtonia*). As a result, the pitcher opening faces toward the rosette center. In other aspects, the basic pitcher morphology of *S. psittacina* is the same as that of other species. The common name is the parrot pitcher plant, from the sideview appearance of the domed pitcher.

The plant blooms in April to May in the Florida Panhandle and their red flowers cover the grassy field amongst many wild flowers. The tall flower scapes grow to 40 cm.

The pitcher, 10-15 cm long, is generally prostrate (lying flat). This renders the general gravity-aided pitfall rather ineffective — and the scheme had to be slightly modified. The prey enters the pitcher through a small opening underneath the domed hood. The round dome is scattered with white areoles. These white windows help light up the otherwise dark dome interior, removing hesitation of prey to venture into the pitcher cavity.

Once inside the dome, the raised orifice column effectively conceals the pitcher mouth from the view — the only true exit out of the trap. The insect is guided toward the light coming from the dome ceiling. The inner surface of the dome is covered with

The cream-yellow umbrella-style creates a sharp contrast against the surrounding red, dangling petals of the flower. In late April, Florida.

" Parrot pitcher plants amend strategies.

fine hairs all pointing in the direction of the pitcher tube. Strategically placed areoles continue to the back side of the pitcher, that guide the prey farther into the pitcher tube.

Downward-pointing retentive hairs which normally occupy only the lower half of the tube in the erect pitchers are much better developed, covering the entire length of the pitcher tube right below the dome. Further, the hairs grow much longer, to a point of intermeshing. This creates a lobster-pot trap, preventing the prey from backtracking its course. Lured by nectar and mislead by bright fenestrations, once an insect has entered the pitcher tube, retreat is not an option. The sharp, spine-like hairs pointing in the direction of the pitcher depths force the victim to move forward. Now, light windows are all but absent. This is a narrow, dark, one-way corridor leading to a dead end, literally.

It is noted that the decumbent pitcher structure (with its lobster-pot trap) is also suited for an aquatic catch, and some water creatures are found caught in a submerged pitcher after occasional flooding. As in *S. minor*, a giant form is known from the Okefenokee Swamp, that produces pitchers to 40 cm long. This variant is formally described as *S. minor* variety *okefenokeensis*.

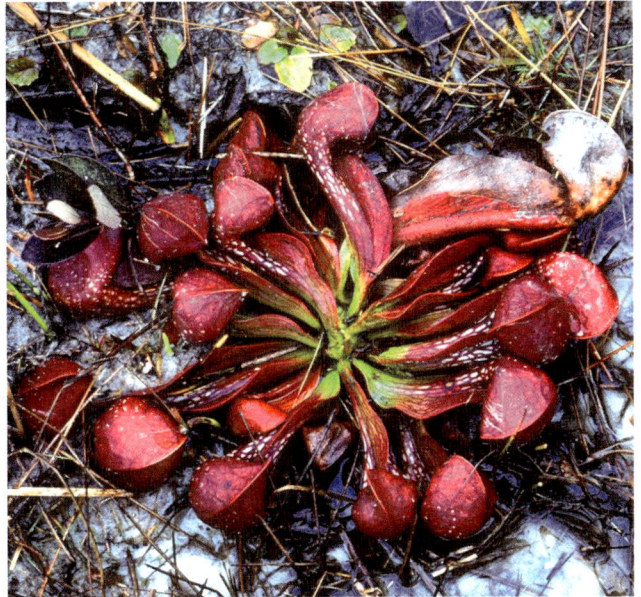

An overwintered rosette of *Sarracenia psittacina*, assuming a deep-red coloration. The substrate is pure-white sandy soil. In Florida, early March.

The longitudinal leaf section of a *Sarracenia psittacina* leaf reveals long, almost intermeshing hairs along the entire length of the pitcher tube. The margin of the well-developed pitcher mouth column rolls up somewhat. Numerous white areoles scattered around the tube entrance provide an unsuspecting visitor the final encouragement to move forward to the point-of-no-return in this "eel-trap" strategy.

85

Flowering parrot pitcher plants sharing a grassy habitat with thread-leaf sundews, *Drosera filiformis*. Bright red blossoms of *Sarracenia psittacina* can be seen throughout the month of May in the Gulf Coastal savannas, long after the flowers of other pitcher plant species are mostly done. In Florida, early May.

TOP: Red *Sarracenia psittacina* flowers among other native wild flowers. In early May, in Florida. MIDDLE: *S. psittacina* often produces spring pitchers that are raised and have a slender tube. In early May, southern Alabama. BOTTOM: Young parrot pitcher plants in Florida, early May.

Sarracenia purpurea

LEFT: *Sarracenia purpurea* ssp. *venosa* in flower, in May, North Carolina. RIGHT: A bug's-eye view of a pitcher of *S. purpurea* ssp. *venosa*. Note the downward-pointing hairs covering the vertical inner wall surface of the pitcher. In July, North Carolina.

This species has the broadest distribution of all in the genus, and two subspecies are recognized: *Sarracenia purpurea* subspecies *purpurea* that occupies the bulk of its huge boreal distribution in the northern range including the Great Lakes region and southeastern Canada, and *S. purpurea* subspecies *venosa* in its southern range from Delaware/Virginia down to the southeastern United States. Subspecies *venosa* is further subdivided into three expressions: variety *burkii*, variety *montana* and variety *venosa*.

The pitcher is bulbous in shape, being decumbent and upward-curved, ranging 15-25 cm in length. A simple (if not rigorous) differentiation of the two subspecies is: Subspecies *purpurea* is tall and slim while subspecies *venosa* is fat-looking, though the two pitcher shapes do look alike. The external surface of the subspecies *venosa* pitcher is far more pubescent and velvet-like to the touch. Also, subspecies *purpurea* (in the north) tends to form a large clump often reaching one meter across whereas subspecies *venosa* (in the south) almost never does. The appearance of a large clump of numerous open pitchers of subspecies *purpurea* is very much reminiscent of a dense colony of the marsh pitcher plant (*Heliamphora*), the South American counterpart.

In the Great Lakes region where the northern form occurs abundantly in bogs and marl fens, the plant is found in a variety of substrates with a pH ranging from 5 to 9. In an acidic habitat, the plant is associated with *Sphagnum* moss, often partially buried in it. If the substrate is alkaline, as in marl fens, the plant is likely to grow with other types of moss (even if *Sphagnum* may be present), or may grow directly in the soil. The plant tends to become larger in *Sphagnum*, whereas in alkaline soils the pitchers are smaller but grow in greater numbers.

The wavy lid of *S. purpurea* is upright and offers no cover for rain. The pitcher gets filled to the rim easily after a heavy rain. However, the plants actively control water volume to an optimal level for insect trapping — to a few centimeters below the rim. The pitcher tends to sit on the substrate due to its curved leaf posture. This provides a stable support for a water-filled pitcher.

Deep red is the general petal color of the flower for this species. In the Southeast, the plant is the earliest to bloom among all pitcher plants. In mid-March, large, pinkish flowers of *S. purpurea* subspecies *venosa* variety *burkii* signal the arrival of spring in the Gulf Coastal savanna. In the far north, the plants flower progressively late in the season. In the south, *S. purpurea* seeds germinate soon after ripening (stratification is not a requirement) whereas for the northern plants, pre-chilling for a duration of one-to-three months in moist condition is almost a necessity. This delays germination until spring, thus averting loss of seedlings by severe winter chill.

In many *Sarracenia* species, pitchers normally die back during winter, but the leaves of *S. purpurea* often persist past the next spring.

Sarracenia purpurea subspecies *venosa*

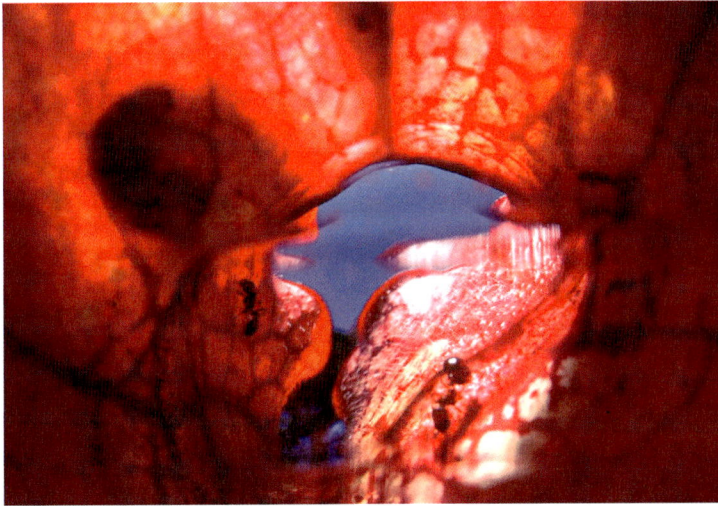

LEFT: The view from the bottom of a pitcher pool, looking up — local ants foraging on the vertical walls of the pitcher lid. In July, North Carolina.
FAR LEFT: A new pitcher of *Sarracenia purpurea* ssp. *venosa* var. *burkii*. In May, in the Florida Panhandle.

IT REMAINS UNKNOWN why this species — and this species alone — has succeeded in extending its territory over such a vast expanse of the North American continent. After the last retreat of continental glaciers (that had covered Canada and northern United States) at the end of the Pleistocene epoch some 10,000 years ago, *S. purpurea* has rapidly spread northward, establishing the current distribution deep into southeastern Canada. Recent DNA sequence comparisons have revealed that *S. purpurea* is sister to all the other taxa in the genus. This suggests that, in *Sarracenia*'s evolutionary history, *S. purpurea* diverged first long before the speciation of all the other southeastern pitcher plants took place.

Sarracenia purpurea subspecies *purpurea*

Freshly caught prey floating in the pitcher of *Sarracenia purpurea* ssp. *purpurea*. In mid-July, northern Michigan. BELOW: The pitchers actively adjust the water level — by absorbing the excess and replenishing the lost water. *S. purpurea* ssp. *purpurea* in northern Michigan, mid-July.

An ant exploring the nectar-baited ridge of the pitcher peristome. The acidic fluid below is well-formulated to take any victim that falls in. *Sarracenia purpurea* ssp. *purpurea* in northern Michigan, mid-July.

91

Sarracenia purpurea
subspecies *purpurea*

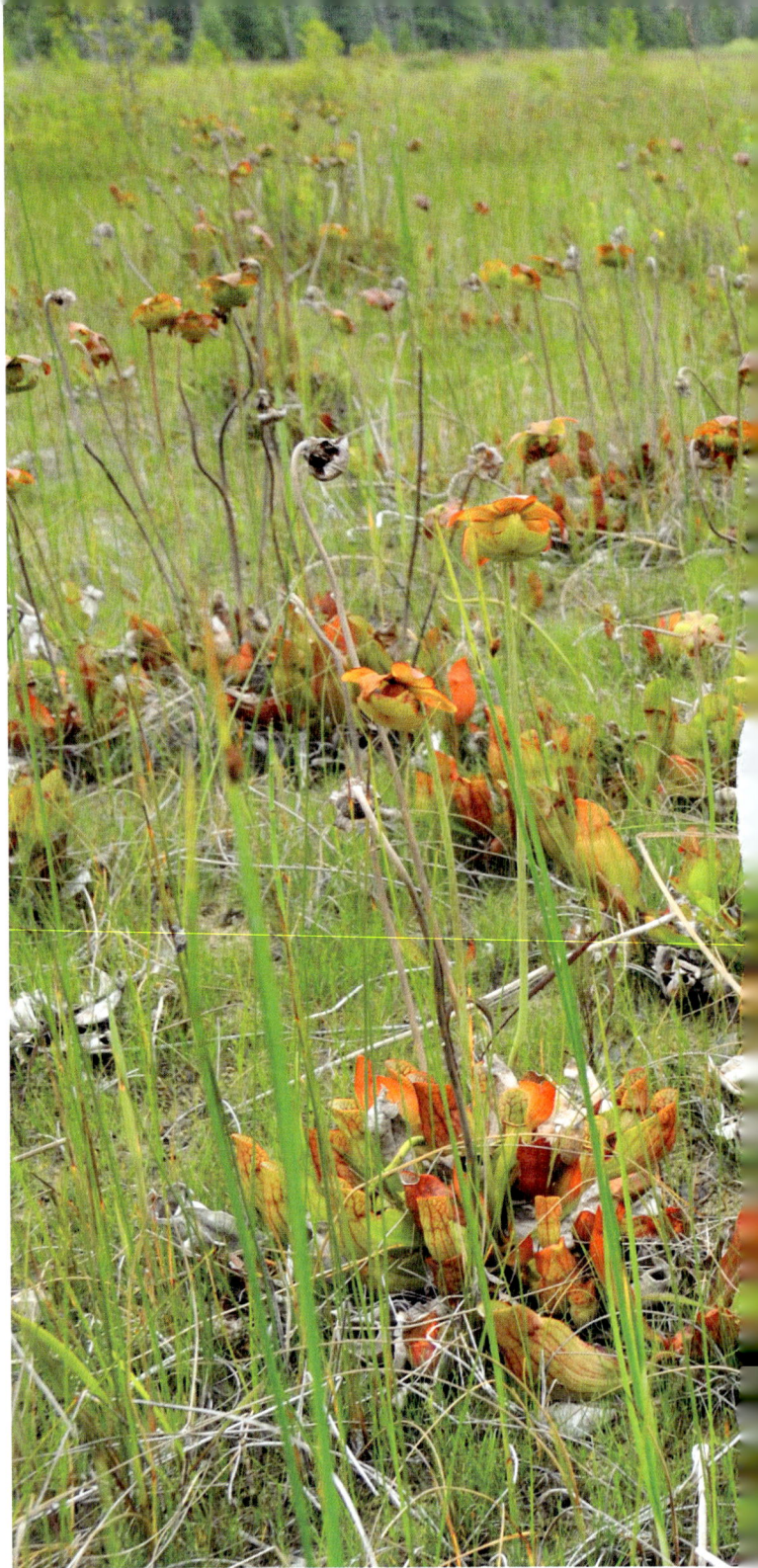

LEFT TOP: A maturing seed capsule of *Sarracenia purpurea* ssp. *purpurea*. In northern Michigan, mid-July. LEFT MIDDLE: Germination of *S. purpurea* ssp. *purpurea*, in mid-July, in a marl fen, northern Michigan. It appears that a mature seed capsule had fallen on the substrate and, after winter stratification, germinated all at once. Many seedlings are showing only twin cotyledons. LEFT BOTTOM: Just a few feet away from the germination above, I found another clump of seedlings, probably one-year old, much further developed with a few tube-like juvenile leaves.

ABOVE: A marl fen along the shore of Lake Huron, harboring an impressive colony of *Sarracenia purpurea* ssp. *purpurea* plants. Other carnivores found in the fen include *Drosera rotundifolia, D. linearis, D. anglica* and *Utricularia cornuta.* In mid-July, northern Michigan. LEFT: Peril awaits even on a non-carnivorous orchid flower. A pollinator (syrphid fly) meets a fatal ambush of a crab spider hiding on a flower of *Calopogon tuberosus,* an orchid commonly seen in the pitcher plant fen in northern Michigan. In mid-July.

Cobra Plant

GENUS *Darlingtonia*
FAMILY **Sarraceniaceae**

The plant was discovered in 1841 by William D. Brackenridge, assistant botanist of the U.S. Exploring Expedition, in a marsh a few miles south of Mt. Shasta in northern California. John Torrey, a distinguished botanist of the nineteenth century,

recognized a close relationship to *Sarracenia*, yet noted a clear difference in floral characteristics, and established a new genus in the pitcher plant family Sarraceniaceae. He named the plant *Darlingtonia californica*, in honor of his friend and botanist, William Darlington. The genus *Darlingtonia* is monotypic, i.e., there is only one species in the genus. The family Sarraceniaceae also includes *Heliamphora,* the marsh pitcher plants from South America.

Recent molecular phylogenetic research indicates that, in the evolution of the pitcher plant family, *Darlingtonia* diverged from the common lineage much earlier than the separation of *Sarracenia* and *Heliamphora*. The basal position of *Darlingtonia* may provide valuable insight into the biogeographic history, suggesting that the ancestral distribution of the genus is where the family Sarraceniaceae originated.

With its restricted present-day distribution, the cobra plant is considered paleoendemic — a surviving relict population — a remnant of what was previously much diverse and widespread ancestors of *Darlingtonia*. The rise of the Rocky Mountains and the eventual formation of the Cascade Range at the end of the Eocene epoch (some 34 million years ago) caused the seasonal drying of a once moist climate throughout the Pacific Northwest. This forced the cobra plant's ancestors to seek refuge in an isolated, climatically stable habitat that we see today.

TOP: A *Darlingtonia* blossom in May, northern California.
OPPOSITE: New pitcher leaves, in July, northern California.

DESCRIPTION

The cobra plant is a herbaceous perennial consisting of a rhizome, with fibrous roots. The tubular pitcher leaves arise directly from the rhizome, forming a rosette. In this species, the hood of a pitcher is well developed to form a dome, with the pitcher opening facing downward. From the frontal edge of the opening grows a two-lobed, fishtail-shaped appendage projecting downward.

A peculiar feature of the plant is that the leaves twist about one half turn as they grow. As a result, all pitchers tend to face away from the rosette center, projecting appendages outward for more effective prey attraction. Based on field observations, the direction of the twist is just about even between clockwise and counter-clockwise, though within a given individual it appears fixed.

In a typical sunny habitat, a mature pitcher leaf stands between 40-60 cm tall. Seen in the field, the overhanging hood of the pitcher — along with a fang-shaped appendage — gives the impression of a deadly cobra poised to strike in imminent defense, hence the common name. Other names for the plant include cobra lily and the California pitcher plant.

Darlingtonia grows in northern California and southwestern Oregon and is endemic to the region. The plants occur in a wide range of altitude. Large colonies are found in montane seeps at an altitude of up to 2400 m, while some scattered populations can be seen a few feet above the sea level in coastal bogs and on lake margins in southern Oregon.

Streams flowing through the mountain meadow create an ideal habitat for this remarkable species of the pitcher plant family. Cold spring waters form a wide band of marshy surface on a mild mountain slope in a sparsely populated coniferous forest. It is in such a setting that the large colonies of the cobra plants thrive.

Many sites are also known to be located on or near magnesium-rich (but nutrient-poor) serpentine outcrops — with ultramafic rock formations containing elevated amounts of chromium and nickel (which may be toxic to plants). These nutritionally poor and phyto-

ABOVE: Stately pitchers stand above the grass surface, shining brightly against the late afternoon light. A mild mountain slope with cold running water creates an ideal habitat for this unique California pitcher plant. In July, southwestern Oregon. LEFT: As the evening sun casts the last rays of the day over a *Darlingtonia* colony, a mysterious glow starts to emanate from the cobra heads covering the meadow in the thousands. In July, southwestern Oregon.

toxic soils provide a competitive advantage for cobra plants possessing a carnivorous lifestyle.

As for accompanying carnivores, cobra plants are often seen with *Drosera rotundifolia* (round-leaved sundew). There are some colonies where *Pinguicula macroceras* (butterwort) grows extensively alongside *Darlingtonia*.

As in the eastern pitcher plant habitats, *Darlingtonia* benefits from periodic, naturally occurring fires. This low-intensity ground fire is said to pose little danger to the established cobra plant populations.

In summer, the daytime temperature of the colonies may reach 35 degrees Celsius while the running water where the roots are submerged rarely exceeds mid-20 degrees Celsius on the same day. This constant supply of cold water seems essential for the healthy and vigorous growth of this species. In nature, the plants are rarely seen in standing water. During the winter months, in contrast, the plants endure extremes of low temperatures in the inland habitats of northern California.

INFLORESCENCE

A flower bud forming in the rosette center during the cold winter months develops into a tall scape (flower stalk) by spring. The solitary, actinomorphic (radially symmetric) flower blooms in a pendulous position at the tip of the scape. The scape may be shorter in statue (30 cm) at the time of flowering, but grows taller in the ensuing weeks to 60 cm. The scape bears ten small bracts scattered along its length.

The dainty flower has five petals, which hang from the base of the dangling flower. The pale-yellow petal is heavily lined with red veins, making the flower appear bright red. The tip of each petal comes together to form a slightly elongated, sphere-shaped corolla. Each petal has a small notch on both sides two thirds of the way down from the attached base. When the corolla sphere is formed, each neighboring notch pair creates one circular opening, five in all, around the lower sphere surface. Five yellow, elongated sepals softly overhang the red corolla. Inside the corolla hangs a large, bell-shaped, five-chambered ovary surrounded by fifteen or so stamens. Underneath the flat bottom of the ovary grows a tiny, five-lobed stigma.

When a flower opens, the anthers shed an ample amount of pollen. A slight tap on the flower causes a large amount of powdery pollen to fall out from the corolla opening where petal tips meet. The anthers continue dehiscing pollen for a few weeks. The expanded bottom of the bell-shaped ovary prevents the stigma from being showered with the flower's own pollen. Also, the flower seems protandrous. That is, the stigma does not become receptive immediately — a common strategy seen in flowers to prevent self-pollination.

In a controlled study, emasculated flowers produced less seeds compared with undisturbed flowers, suggesting

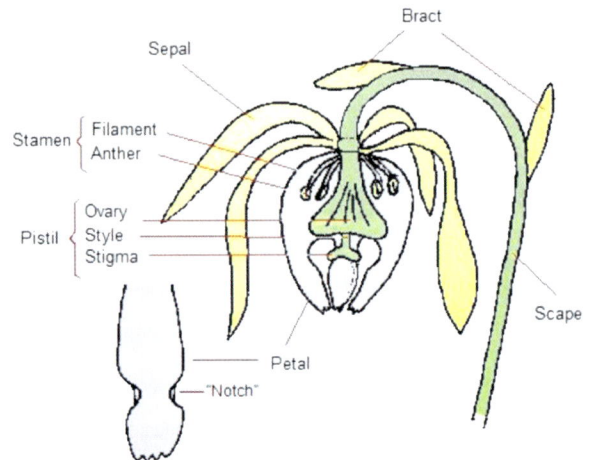

Cobra Plant Flower

ABOVE: A flower bud on a crooked scape. A few days before opening, the five sepals tightly fold the nodding, spherical corolla. A small spider anxiously awaits behind one of the bracts on the scape. In early May, northern California. LEFT: Towering inflorescences of *Darlingtonia*. The deep-yellow sepals softly hang over the brightly red spherical corolla, creating a vivid visual contrast. The mid-afternoon sunlight provides a perfect illumination for this eerie-looking, colorful creation of nature. In late May, northern California.

self-pollination is also contributing to seed production in nature — though we cannot dismiss the possibility that manually removing pollen-producing anthers from the flower inevitably damages the sensitive *Darlingtonia* flowers in a way that negatively affects seed crops.

THE FLOWERING PERIOD VARIES slightly depending on altitude. In southwestern Oregon, the blossom starts in mid-to-late April, and continues into the months of May and early June. In northern California, the blossom typically occurs a few weeks later. In northern California's inland habitats, flowering starts in early May and lasts for a few weeks. Petals remain on the flower for another week or so. Within the same habitat, shaded plants delay their flowering slightly.

The *Darlingtonia* flower has a rather distinct scent, something reminiscent of green vegetation freshly cut. It is a pleasant aroma, one of perfume fragrance, and is fairly strong for a newly opened flower.

Anthocyanin-Free Variant A population of anthocyanin-free plants has been found in California in 1994. The flower has no red venation and is totally yellow. The pitcher also lacks any red pigmentation and remains all-green to yellow. The plants are identical to the normal form in any other respect. This variant has since been formally described as *Darlingtonia californica* form *viridiflora*.

The *Darlingtonia* flower, with one sepal and two petals removed to show the corolla interior. Many stamens surround the bell-shaped ovary. The five-lobed stigma protrudes on the bottom of the ovary.

IN *DARLINGTONIA*, THE COLOR SCHEME appears similar between flowers and traps. Are they competing for the same class of visitors?

The flowers appear in the spring, often in a wholesale explosion of blossoms throughout the colony. At the time of flowering, new spring pitchers are just emerging from the rosette center, which become fully mature and functional in several weeks, long after the fertilization is completed. This provides a clear margin of safety for pollinators. In mild climates, leaves of the previous season may persist into spring. While this may present an annoyance for pollinators, key attractants such as nectar and bright colors are not present. Also, the old pitchers may be tilted, making a gravity-based pitfall trap ineffective.

Cobra plants produce flowers long before the new spring pitchers for pollinator safety.

LEFT: A fine stand of cobra plants in northern California, early May. The plants are growing in a very wet condition in this colony, and the plants are partially submerged in a cold, flowing water. Numerous blossoms are soon to be followed by a rush of new leaves, signaling the onset of the trapping season in cobra plant country. BELOW: A pair of spring leaves shooting up from the rosette center of the plant submerged in cool, spring water. These very first pitchers are the tallest of all pitchers to follow. In mid-May, northern California.

The early-morning rays of the sun pierce through the colorful blossoms of *Darlingtonia*, creating a spectacular floral display. Morning dew drops still remain on the yellow sepals that overhang the bright-red, spherical corollas. In mid-May, northern California.

Fully matured pitchers (40-60 cm) of cobra plants crowd a grass-covered field, in early July, southwestern Oregon. A cold mountain spring creates a marshy surface on a mild, south-facing mountain slope, providing a constant supply of fresh water to the root of the cobra plants. These pitchers are the very first pitchers of the season — and the tallest of all pitchers to follow. Numerous fruit capsules are seen at the tip of the tall scapes. Evenly spaced, small bracts along the scape — which were more clustered at the top at the time of flowering, two months earlier — indicate the scapes have fully stretched to their maximum (85-100 cm) after fertilization. It will take another month or two for seeds to mature.

TRAP STRUCTURE AND ATTRACTION

The pitcher appears light yellow-green on the domed hood, becoming darker green toward the base. The areas around the pitcher opening and the fishtail appendage assume varying degrees of a reddish coloration in plants growing in full sun.

The basic trap scheme is the same as that of the eastern pitcher plants. Together with the brilliant colors of the leaf, prey are lured to the trap by nectar-secreting glands scattered over much of the pitcher exterior. These nectaries form nectar trails that direct crawling insects to the pitcher opening.

The fishtail appendage provides a convenient landing site for flying insects. It is a perfect feeding ground as well, for the nectar secretions are abundant. The upper surface of the appendage is lined with short, stiff hairs all pointing toward the pitcher orifice. This encourages the insect to ascend ever closer to the trap entrance overhead.

ABOVE: A bug's-eye view of the pitcher dome interior, looking down towards the orifice. RIGHT TOP: Looking toward the dome front. RIGHT BOTTOM: Looking down into the narrowing pitcher tube. The slippery inner surface ensures a smooth ride deep into the spiral pitfall trap.

OPPOSITE: The fishtail appendage that provides a convenient landing ramp for a winged insect is but a mere stopover before proceeding into the brightly-lit cobra dome. In August, northern California.

MUCH OF THE UPPER PITCHER, particularly on the dome, is scattered with many areoles, called fenestrations. These patches are completely void of chlorophyll and other pigments, forming truly translucent windows. Seen against the bright sky, the well-lit ceiling of the dome persuades an insect foraging on the appendage to venture into the dome interior.

The edge of the orifice rolls inward, forming a "nectar roll" around the pitcher opening. This is where the nectar secretions are the heaviest. Once inside, the chance of escape becomes slim. The rolled-up margin of the orifice somewhat conceals the true exit from the insect's view, and the light windows — more numerous toward the pitcher back — provide an illusion of a gateway to freedom. An insect, instinctively seeking an exit in the direction of light, often slams against the dome ceiling in its attempt to fly through, and tumbles down into the spiral depths of the cobra leaf. For those wishing to tread, the dome ceiling is lined with short, stiff hairs (1 mm) all aligned toward the back of the dome — just above the spiral tube.

> **Bright windows create deceptive exit to freedom.**

The inner pitcher tube walls right below the dome are hairless and are covered with detachable waxy scales, creating poor footing for the insect. Further down along the increasingly narrow pitcher tube grow long, downward-pointing retentive hairs that forbid the ascent of fallen insects. This retention zone continues to the bottom of the pitcher. The prey drowns in the liquid accumulated at the base.

ABOVE: Numerous translucent fenestrations (areoles) brighten the dome interior. The ceiling of the pitcher dome is also lined with short hairs (1 mm long) all pointing in the direction of the spiral tube.

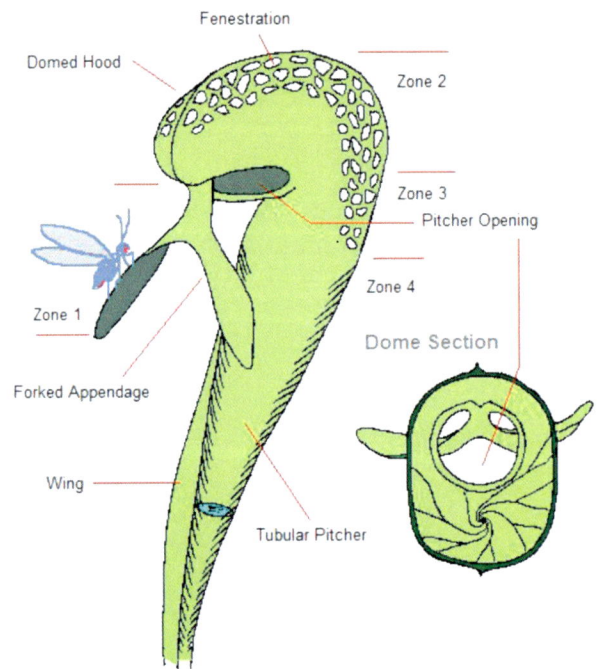

Fenestration
Domed Hood
Zone 2
Zone 3
Pitcher Opening
Zone 4
Zone 1
Dome Section
Forked Appendage
Wing
Tubular Pitcher

Cobra Plant Leaf

108

DIGESTION

> **"** *Darlingtonia* does not expend energy for enzyme production — ever-ubiquitous commensals will do the job.

The forked appendage, seen in the air here, often touches the ground for a short pitcher, serving as a ladder for crawling creatures. In July, northern California.

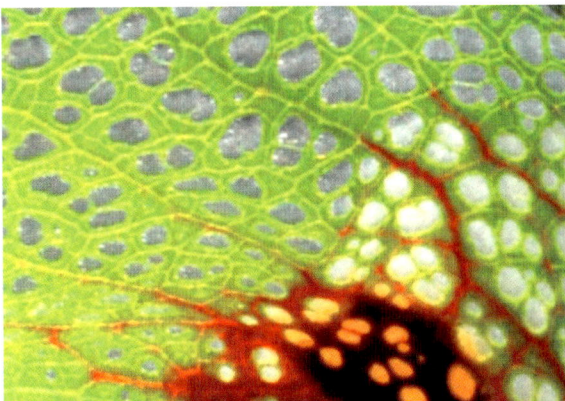

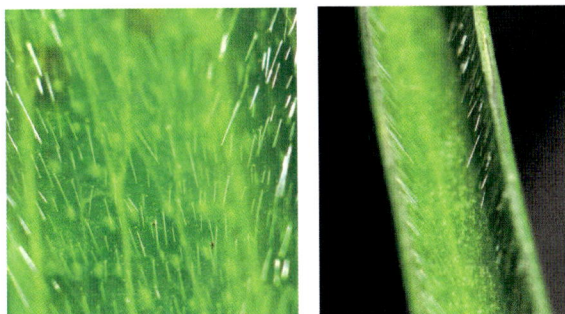

The overhanging domed structure of the leaf all but precludes the possibility of rainwater entering the pitcher. In *Darlingtonia*, an unopened pitcher already contains some liquid. Unlike most species of the pitcher plants in the East, the cobra plant does not possess any digestive glands in the pitcher walls and no enzyme secretions have been detected. The digestion of prey is carried out solely with the aid of externally introduced bacteria and some commensal organisms.

In the pitcher fluid, many white worms are observed feeding on captured prey. They are larvae of a fly (*Metriocnemus edwardsi*). Small fragments of the decomposed prey are then consumed by mites (such as *Sarraceniopus darlingtoniae*). As in the *Sarracenia* species, the pitcher fluid of *Darlingtonia* forms a micro-ecosystem of inquilines.

In spite of the lack of enzyme production on the part of cobra plants, the trap exhibits highly advanced adaptations to carnivory in its morphology as well as physiology. Studies have shown that a certain chemical stimulation (like a small piece of meat thrown in the pitcher) precipitates copious secretions of liquid into the pitcher.

The bottom part of the pitcher, where the liquid is retained and the digestion takes place, does not possess any special glands. The permeability of the inner wall due to cuticular discontinuity allows the absorption of digestion products into the leaf tissue.

LEFT TOP: The truly translucent and cloudy dome fenestrations; in Zone 3, which starts around the lower end of the dome, the inner hairs abruptly become absent and the fenestrations turn white and cloudy. LEFT MIDDLE TWO: The upper and lower (narrower) part of Zone 4 with downward-pointing hairs. LEFT BOTTOM: Drowned prey in the pitcher liquid. In northern California, July.

ABOVE: Dense growth of cobra plants on a mountain slope in Oregon, in July. The domed cobra heads glow in the late-afternoon light, enhancing a surrealistic mood of these pitcher plants of the Pacific Northwest. RIGHT: A huge pitcher reaching 60 cm in height, which is just about the maximum size attainable in the wild (without etiolation) on a healthy plant growing in an ideal, sunny habitat. Note the distinguished "mustache" appendage of this pitcher — albeit a bit uneven, sloppy paint job. In July, southwestern Oregon.

REBECCA M. AUSTIN

" Her passion, dedication, toward *Darlingtonia.*

Moving to northern California with her husband and their three young children, Rebecca Austin was fascinated by this unique California pitcher plant. Living near large cobra plant sites, including Butterfly Valley (now a protected nature preserve), she spent many days observing the plants in the field, sometimes carrying her sewing to the colony, and even setting up a tent for overnight observations. She pursued her study of the plants, even in the midst of a violent thunderstorm, convincing herself that the rainwater did not enter the pitcher.

Throughout the recorded history of *Darlingtonia*, Rebecca Austin was the first to have investigated the plants in such detail. From 1875 to 1877, she diligently communicated her observations and findings to W. M. Canby, a botanist in the East. Canby sent stationery to help her work and encouraged her observations. The then-newly-published book *Insectivorous Plants* by Charles Darwin was sent to her to assist in her studies. Rebecca Austin had already noted that the plants required plenty of cold water for healthy growth.

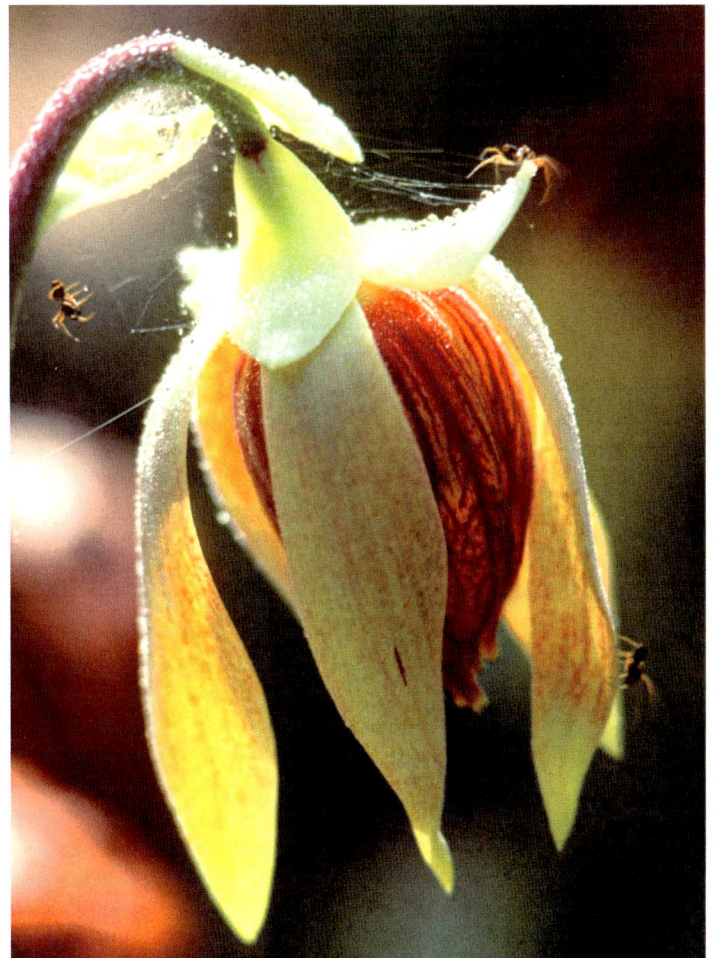

Construction work starts early at day-break. On a freshly opened flower still covered with morning dews, small spiders busily spin webs with precision. Emitting a thick strand of silk in the wind, spiders can "fly" from one flower to another with utmost ease, possibly transporting pollen in the process. In early May, northern California.

In search of
ELUSIVE POLLINATORS

"In spite of the past 170 years of study since the discovery of *Darlingtonia californica*, the pollination biologist is yet to identify the major pollinator of cobra plant flowers. The unique flower morphology appears to suggest a highly evolved floral adaptation in view of its reproductive strategy. There is no firm field observation to date, however, that points to the existence of major pollinators of the flower.

A spider taking up residence in a *Darlingtonia* corolla. In early May, northern California.

Researchers have been trying to identify the pollinators of cobra plant flowers — with limited success to date. It is a common observation for those who have visited any colony in September that countless flower scapes stand in the field holding fruit capsules filled with thousands of viable seeds. Whoever is responsible for the abundant crop of seeds, the unidentified pollinator has eluded discovery for the past 170 years.

Rebecca Austin, through her keen observations, suspected the ever-ubiquitous spiders to be the pollinator of cobra plants. Many observers since Austin have witnessed spiders and their silky webs on the cobra plant flowers. In fact, it is often difficult to find a flower totally free of spider webs. This observation holds true for both Oregon and California habitats.

IN A RECENT STUDY CONDUCTED in Oregon, only a few sightings of insect visitations have been recorded during daytime observations. Of nearly 1800 insects captured for analysis around the *Darlingtonia* colony in the months of April through June, only 8 individuals (4 species) were found to carry *Darlingtonia* pollen. Insects either directly observed visiting the blossoms or caught carrying *Darlingtonia* pollen include one unidentified dipteran, several flies (Conopidae and Tachinidae) and a few beetles (Buprestidae and Cleridae). During the same period, 85 spiders (14 species in 10 families, with *Clubiona* and *Theridior* being the majority)

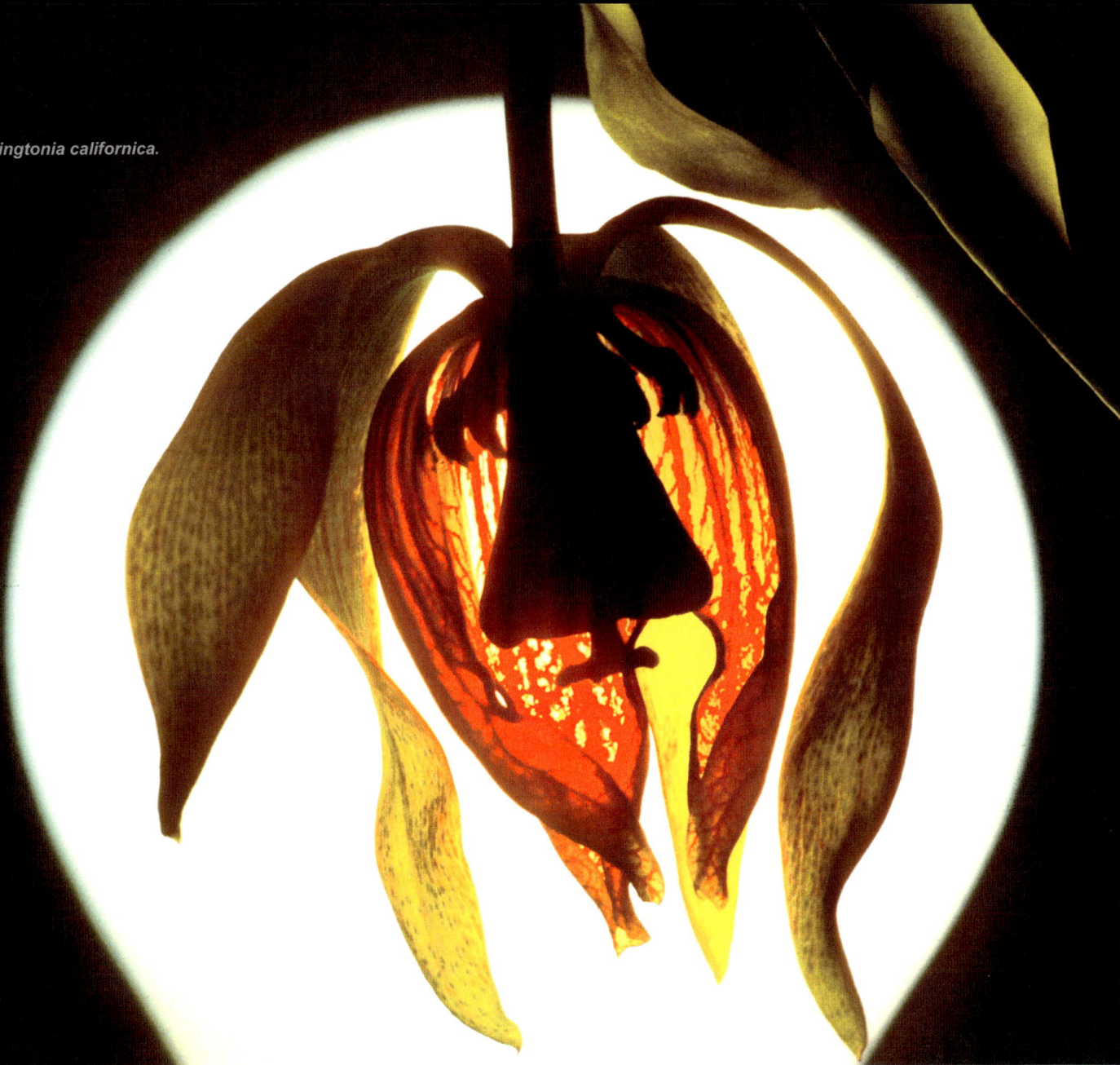

Darlingtonia californica.

" Will the real pollinator please stand up?

observed in and around the flowers were collected and analyzed — with 79 of them carrying *Darlingtonia* pollen.

Could spiders be a *legitimate* agent of pollination for which the flower has evolved over millennia?

Yet in other studies in California, repeated floral visitations by multiple insects have been noted by different observers. A solitary bee, *Andrena nigrihirta*, and, to a lesser degree, a European honey bee, *Apis mellifera*, are emerging as strong suspects in this pursuit — although several other insect species caught in flight with *Darlingtonia* pollen are just as strong a candidate for pollinators.

As for the pollinator riddle, these excellent field studies

— so painstakingly undertaken by dedicated field botanists — seem to confirm, if anything, that indeed the pollination by winged insects is dauntingly rare for *Darlingtonia*. These observations may also indicate different pollinator communities for different sites in Oregon and California.

According to one of these California studies, the visitation frequency by *Andrena nigrihirta*, per flower, is estimated to be somewhere in the range of one visit every other day to about one visit every ten days, depending on the observation sites.

It would seem *Darlingtonia*'s reproductive strategy ought to rely on something more predictable. Do the spiders provide a safety net for this uncertainty?

113

WHILE SPIDERS ARE NOT GENERALLY associated with pollination, one cannot deny the fact that the physical presence of spiders in California and Oregon, on virtually all flowers, in and of itself, represents a valid pollen vector. Since spiders are observed moving among flowers, cross pollination is a definite possibility though selfing may be more probable.

When it comes to cobra plant flowers, however, delicate sepals are twisted by the web, bracts tied to the flower stem. Webs are often spun in the flower interior, with the stamens bundled together. On some occasions, the stigma is completely covered by the web.

Many field observers may tend to agree that this is a case of massive spider infestation! Virtually all flowers in the field are occupied by spiders. It is way too risky for insects to enter the flower. Yet-to-be-identified pollinators are prevented from servicing the flower because of the uninvited arachnid visitors.

Is the fertilization already accomplished by the time spiders commandeer the flower? Observations in the field show that the flowers are occupied by the spiders at the outset. When the colony is about to bloom, spiders are already waiting in the wings ready to attack. It takes only a week for the army of arachnid solders to construct silken networks over the entire colony.

" Spiders, spiders, … and more spiders.

In the absence of other creatures transporting pollen — at least not in sufficient frequency — must we conclude that spiders are indeed the major pollinator of the cobra plant? This is a hard-to-swallow conclusion, considering the unique morphology of the flower.

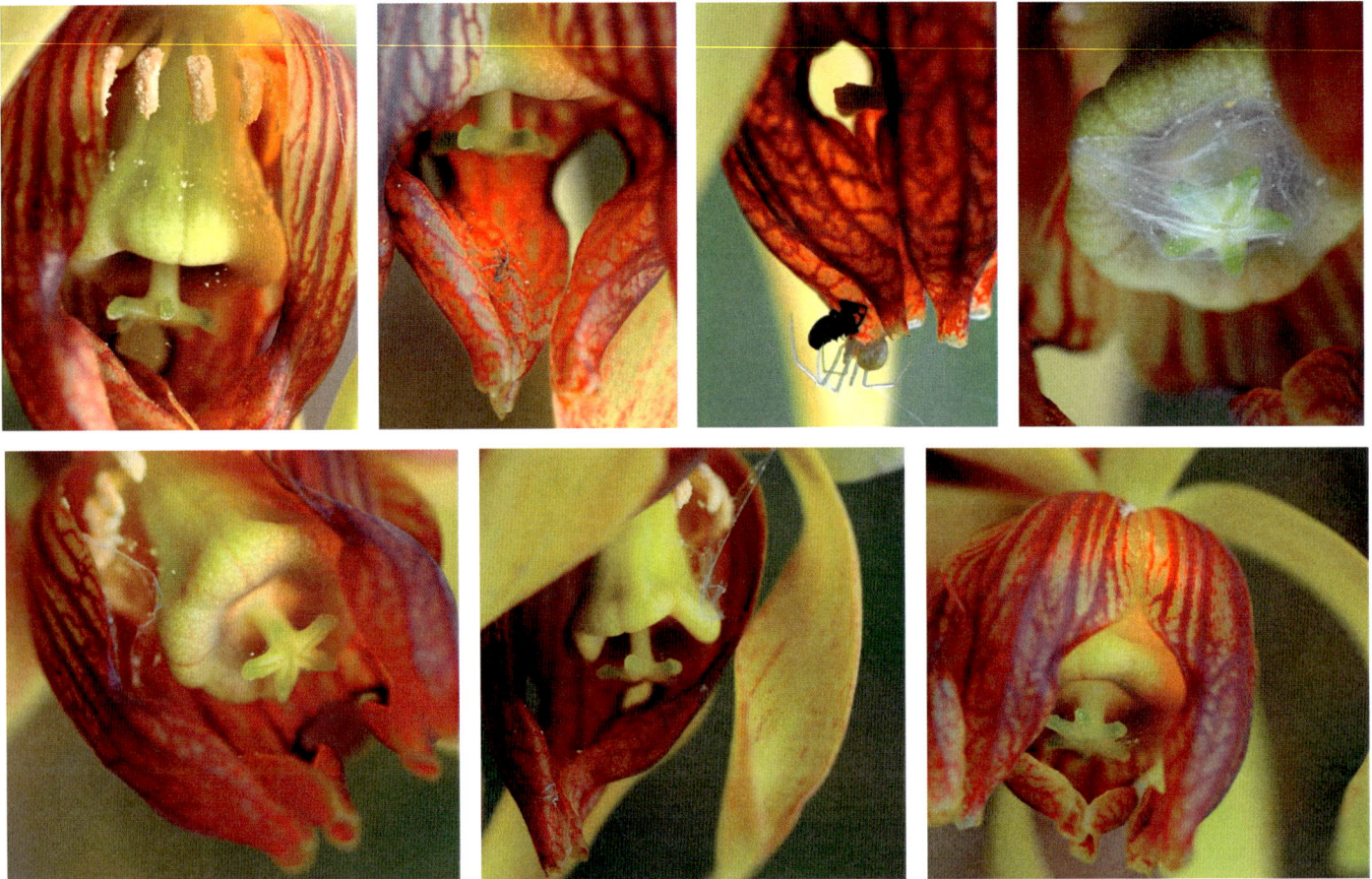

Spiders and their webs on the flower, in May, northern California. Webs spun in the flower interior are dusted with pollen. Observations indicate some arachnid species may be utilizing nectar and/or pollen for their dietary supplement.

Cobra plant blossoms, in May, northern California. High on a tall scape and well protected against heavy rain, a spherical corolla chamber affords a perfect hideaway — with a view — for an uninvited guest.

An Early Bloomer of the Forest Given the limited supply of pollinators in the mountain, massive blooming of *Darlingtonia* in May in northern California is interpreted to be an effective strategy to avoid competition with other wild flowers.

At the height of the cobra plant blossoms, many flower scapes cover the field — with relatively few other flowers to steal the attention from them — and yet, a few flying insects hovering over the field seem largely indifferent to these colorful blossoms. This is true, in spite of *Darlingtonia*'s best intention to accommodate a safe and friendly visitation — high flower positioning, a sweet fragrance and ample pollen offerings.

ABOVE: A seductive stare often goes ignored by flying insects hovering over the field. There is nothing to indicate this flower structure is aimed at a specific pollinator. Some believe a bee to be the yet-unidentified pollinator for the cobra plant flowers while some point out the total lack of investigation for nocturnal creatures. In early May, northern California.

Pollination Strategy *Darlingtonia* flowers are eye-catching. The flower exhibits vibrant red petals and is sweetly fragrant with abundant pollen — features commonly associated with bee pollination.

What is the true purpose of this flower structure? If a *Darlingtonia* pollinator is a generalist visiting other "normal-looking" flowers, what sort of evolutionary pressure necessitated the creation of this extravagant floral masterpiece?

An insect pollinator enters the corolla through one of the five circular openings. The five-lobed stigma projecting under the bell-shaped ovary is located at the same level as the circular openings. This causes the pollinator to brush the stigma immediately upon entry and deposit the pollen collected from the previously visited flowers. As the insect ascends inside the corolla in search of nectar, it now collects pollen of that flower. When the insect is ready to leave, it does so by sliding down the ovary slope. There is a good chance the pollinator will exit the flower through one of the circular openings without touching the stigma again, which is somewhat hidden under the expanded bottom of the ovary. This structure seems to discourage self-pollination. In the event the flower's own pollen was deposited on the stigma upon the pollinator's exit, well, the fertilization process might have already been started minutes earlier — or self-pollination may occur, which is not the end of the world.

THE FLOWER MORPHOLOGY does not seem to favor any particular class of creatures for pollination. This is supported by the observations of multiple insects carrying *Darlingtonia* pollen. Floral visitations by the European honeybee (*Apis mellifera*) further corroborates the point that any insect — even a newcomer — can readily assume the role of a *Darlingtonia* pollinator as long as the "shoe fits." Note that the circular openings of the corolla formed by indentations of the adjacent petals are merely a visual cue for a pollinator and do not limit the insect size, for a larger insect can easily push the petals apart and gain access to the corolla interior to effect pollination. In conclusion, given the overall size of the flower (2 cm across) and the circular openings (0.5 cm across), any insect the size of a normal housefly will fit the bill.

LEFT: Pollen grains cling on the tip of the five-lobed stigma projecting at the bottom of the bell-shaped ovary. FAR LEFT: Anther dehiscence (pollen shedding) starts immediately after the flower opens but stigmatic exudate is not present initially — a common strategy in flowers to discourage self-pollination.

" Is the *Darlingtonia*'s unique floral structure a mere legacy of evolution?

EVEN IF AN INSECT IS THE TRUE pollinator, given the generally observed circumstances where spider webs are visibly spun over flowers, it is unlikely that a normal pollination scenario will play out. It is a suicidal mission for an insect to enter the flower — a bee must be out of his mind or insanely hungry for pollen.

Could it be that the winged insect pollinators have been chased away by the massive invasion of spiders — in space and time — and, as it turned out, the plants are content with this arachnidan arrangement, as evidenced by the healthy harvest of seeds?

It is not inconceivable that the unknown pollinator had become extinct (for some time), leaving the flower without a partner. Floral adaptation requires tens of thousands of years to take effect; extinction of a species can occur in a geologic eyeblink. Is the seemingly contrived floral structure of *Darlingtonia* a mere legacy of evolution?

Given that the pollinator-flower relationship is not something carved in stone, who will be attending *Darlingtonia*'s flowers a million years from now is anyone's guess.

Silver threads of silk blowing in a spring breeze gently remind us of an ever-perplexing and interwoven fabric of nature.

The search for the elusive agent of pollination continues.

As evening draws to a close, a cheerful dance of bees and butterflies gives way to a world of creatures of the dark. Moths fly over the grassy field, spiders emerge from hiding, and buzzing crane flies hover over water in a mating swarm. Cold air begins to envelope the serenity of the forest as silent blossoms of the cobra plants continue into the dead of night. Are nocturnal pollinators on their way for a midnight feast? In northern California, early May.

Towering cobra leaves in a montane habitat in southwestern Oregon, in July. The bright sunlight pierces through the foliage, revealing the intricate network of colorful venations that cover the domed pitcher leaves.

PITCHER LEAVES

" **Cobra plants are said to be a compass plant.**

In *Darlingtonia,* all adult leaves are not the same height. On a healthy, mature plant growing in an ideal sunny habitat, the leaf size varies greatly from a full 40-60 cm at the tallest end to less than several cm. Unlike the pitcher plants in the East which produce uniform-sized adult leaves, the cobra plant produces progressively smaller leaves during its yearly growth cycle. The very first pitcher to emerge right after flowering is the tallest, and the next one the second tallest.

It is Rebecca Austin who first noted the compass nature of the plant. She observed that the plant always produced pitcher leaves in pairs, a total of 10-18 leaves per year. The first two large leaves in early summer would orient in the north-south direction, the next two in the east-west direction. The first four pitchers thus point in the direction of the four major axes of a compass. Some others have confirmed her observations. Field examination corroborates this tendency — though with occasional *violations* to the rule. It ought to be noted that determining the leaf direction is not always easy in the field due to the plants' dense growth, and also because of the horizontally-growing rhizome from the tip of which new pitchers emanate.

REPRODUCTION

After fertilization, a tall scape grows further as it straightens itself from hooked to erect posture, often reaching 100 cm. In nature, seeds are set by September. Each ripe fruit capsule may contain as many as 1000-2000 seeds. A characteristic seed has numerous tiny projections for animal dispersal. Floating seeds in the stream may find new sites to colonize along the water path.

Darlingtonia seeds germinate better after stratification. This is a period of cold temperatures in a damp environment. In nature, the seeds delay germination until spring to avoid the loss of new seedlings during the winter cold.

The cobra plant is a slow grower. It takes two to three years for juvenile, tubular, pointed-end leaves to assume the characteristics of a mature pitcher. A few more years are required for the plant to flower. The plant continues to grow, producing larger leaves every ensuing year until it reaches its size maturity in seven to ten years.

Asexual Reproduction The cobra plant exhibits vigorous vegetative reproduction. A mature plant often produces another growth point (crown division) in the rosette center, eventually growing into two plants. In addition, the plant habitually produces long underground runners, or stolons, often reaching a meter or more in length. The tip of a stolon develops into a new plant. This ability of the plant to reproduce asexually often results in rather dense growth characteristics, as typically seen in the wild cobra plant populations.

ABOVE: A maturing fruit capsule, in August, northern California. By seed set, the flower scape achieves its full length of one meter or so for effective seed dispersal. LEFT: Vegetative reproduction — a new plant is formed at the tip of a long stolon connected to the mother plant. In May, northern California.

One-year-old seedling growing
in a southern Oregon mountain
seep, in August.

RECAPITULATION

Recent molecular phylogenetic analyses show that *Darlingtonia* is the basal genus in the family Sarraceniaceae. The term "basal" means the branch leading to the present-day *Darlingtonia* diverged first from the common ancestor of all three genera. The term sometimes carries misleading connotations of being "primitive." While greater diversification on the non-basal branch (leading to *Sarracenia* and *Heliamphora*) may imply more evolutionary changes, the fact remains that all three genera received the same geologic time to reach the form we see today. To judge which is more primitive, we must analyze the extant plants. Based on morphological as well as physiological evidence, it does appear *Darlingtonia* is the most primitive of the three.

The juvenile leaves of the cobra plant are much simpler tubular shape. The pointed tip of the leaf is a precursor of the fishtail appendage of the adult pitcher.

THE THEORY OF RECAPITULATION is a biological hypothesis that, in the development of embryos, organisms go through stages that reflect the evolutionary history of their remote past. It is like viewing an extremely condensed version of a movie of their millions of years of evolution.

If this holds true in *Darlingtonia* — even to a small degree — the pointed-tip tubular leaf of the seedling may give us a glimpse of what the cobra plant's long-gone ancestors might have looked like, and reveal, for that matter, some clues about the progenitor of the pitcher plant family itself.

" **Ontogeny recapitulates phylogeny.**

TOP: The pointed tip of a red juvenile leaf. RIGHT: Germination of a seed, with the embryonic root. FAR-RIGHT TOP: After stratification, the seeds germinated all at once — 1000-plus in all — in a crowded seed capsule that failed to deploy properly and fell straight on the moist ground. In mid-May, northern California. FAR-RIGHT BOTTOM: Seedlings with twin cotyledons (often three or four of them).

OPPOSITE: *Darlingtonia* seeds, measuring 2 mm in length. The seeds are covered with numerous projections for animal dispersal.

DORMANCY

In contrast to its intolerance to the summer heat, *Darlingtonia* endures the cold temperatures quite comfortably in the inland habitats of northern California. While the populations along coastal Oregon enjoy a rather mild climate throughout the winter months, the temperature in the montane habitats readily drops below freezing in winter. The pitchers of the cobra plants are often seen covered with frost, ice, and snow in these severe conditions — yet the foliage maintains its green color to continue photosynthesis.

" Winter envelopes cobra plant country.

OPPOSITE & ABOVE: Pitchers of cobra plants covered with frost, at daybreak, on a cold winter day. Note that the pitchers remain green, suggesting photosynthesis is continuing. In northern California, early December.

Sundews

GENUS *Drosera*
FAMILY **Droseraceae**

Glistening in the sun like a cluster of diamonds, the name "sundew" aptly describes the magical beauty and charm of these adhesive-trap carnivorous plants that ensnare

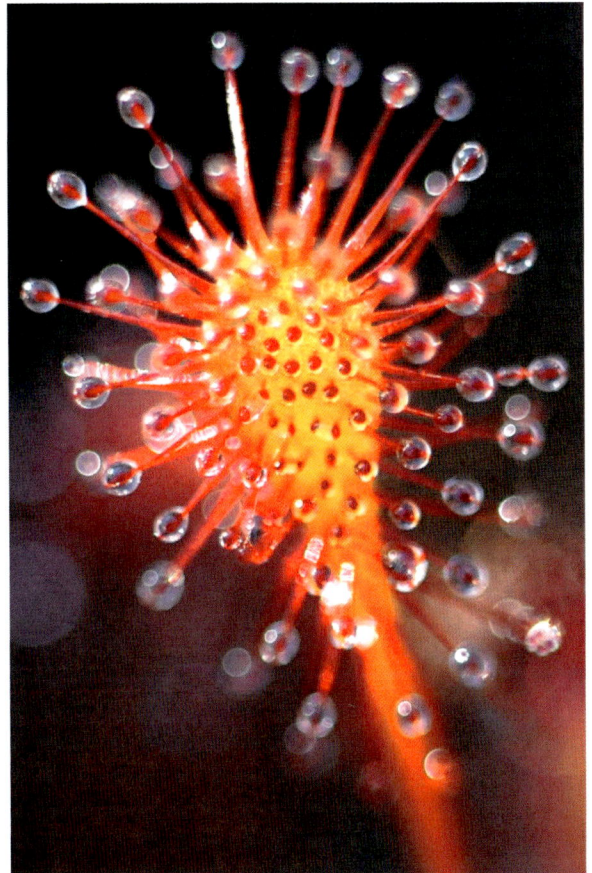

small animal prey with their innocuous-looking dews. The genus name *Drosera* is derived from a Greek word *droseros* for "dewy."

Drosera is among the largest genera of carnivorous plants along with bladderworts (*Utricularia*). There are well over 200 species of sundews worldwide, both in the southern and northern hemispheres. By far, the heaviest concentration of the species occurs in Australia — particularly the southwestern part of Western Australia — with nearly fifty endemic species in this continent alone. The southern part of Africa, including the Cape Floristic Region, is also known for high endemism, followed closely by South America.

Although the genus as a whole has a vast boreal distribution throughout much of Europe, only three species (*Drosera rotundifolia, D. anglica* and *D. intermedia*) are found in Europe. The North American continent harbors seven species, of which three are common with Europe. Two species (*D. linearis* and *D. filiformis*) are endemic to North America.

TOP: A dewy leaf of *Drosera intermedia*, early morning in mid-July, southern Michigan.
OPPOSITE: An uncoiling new leaf (inward circination) of *D. filiformis* var. *tracyi*. In early May, Florida.

DESCRIPTION

“ **Glistening dews are a deadly trap in disguise.**

Sundews are perennial or annual herbaceous plants. Many species form a rosette of leaves, lying low on the ground, with a short stem and fibrous roots. Others have long stems, some exceeding two meters in length. Sundew leaves vary in size and shape among different species, but the basic structure remains the same for all sundews. The leaf is of two parts: a narrow petiole (leaf stem) and a leaf blade modified into an adhesive trap. Stipules (small appendages) at the base of the petiole are sometimes useful for taxonomic identification in some species.

Sundews typically grow in permanently moist peaty soils and other wet areas. In Australia, the habitats may experience a dry summer. As expected from the large genus size, sundews are a diverse group of plants that come in various growth characteristics.

Some temperate species form winter hibernacula. The *Drosera* hibernaculum is made up of tightly packed bud-like leaves formed in the rosette center that withstands cold and desiccation.

Some have adapted to a dry climate. There are forty-some Australian species that develop underground tubers. During the dry summer season, all leaves above the ground die out. The tuber re-sprouts in the fall when the moisture returns. There are some thirty species also from Australia collectively known as "Pygmy" sundews because of their miniature size (typically less than 15 mm in rosette diameter). They are characterized by gemmae production in the rosette center. The gemmae are tiny, scale-like organs that sprout to become a new plant. In addition, there are some sub-tropical species from Africa and Australia.

TOP: A typical habitat of *Drosera anglica* in a sub-alpine swamp in southern Oregon, early July.
LEFT: Thread-leaf sundews (*Drosera filiformis* var. *tracyi*) blooming in the early morning light. Note that all flowers are facing in the southeasterly direction. In the Florida Panhandle, early May.

INFLORESCENCE

Sundew flowers are predominantly white to light pink, and small, but some species do produce large, colorful flowers of yellow, red and orange petals. The flowers are actinomorphic (radially symmetric). The general scheme for sundew flowers is parts of five: five sepals and five petals. Besides occasional deviations in any species, there are some steady departures from this norm. One pygmy sundew, *Drosera pygmaea*, typically produces four-petaled flowers (though five-petaled flowers are not uncommon). The flowers of *D. heterophylla* often have 7-10 petals. Many sundew flowers also have five stamens.

Most sundew flowers are believed to be insect-pollinated. The physical separation of flower and trap achieved by a tall flower stalk in many species affords the pollinator safety during pollination. Some sundews produce their flowers amid a forest of glandular leaves — as if the flowers themselves are deployed to lure the prey, by design. Studies of some sundews suggest, however, that the flowers appear to attract a distinctly different class of insects than the prey trapped on the leaves, and the overlap of insect species found as both pollinator and victim is less than a few percent of the typical prey population.

Usually the flower opens for one day for the majority of species. In many temperate sundews, if the pollination does not take place, the anther and stigma are brought together as the flower closes to effect self-pollination.

White flowers of linear-leaved sundews (*Drosera linearis*) blooming amid glandular leaves. Navigational skill is of paramount import for pollinators. In northern Michigan, mid-June.

Round-leaved sundews (*Drosera rotundifolia*) showing vigorous growth in a water-logged peaty substrate in a northern California seep. In early July.

ADHESIVE TRAP

" The tentacles play a crucial role in the whole process of prey trapping and the subsequent digestion.

LEFT: The round-leaved sundew, *Drosera rotundifolia*, entrapping a hoverfly. Bee-color mimicry, no doubt, has contributed to the welfare and survival of the hoverfly — only to be caught by a sundew who doesn't give a hoot about the skin color of its food. In early May, northern California. OPPOSITE: *D. rotundifolia* capturing a bug. Many marginal tentacles have already bent towards the center of the leaf blade to secure the prey. In a northern California seep, late May.

The leaf blade of sundews is covered with fine hairs each tipped with a sticky glue. These glandular hairs are called tentacles. They grow mainly on the adaxial (upper) surface of the leaf blade but are also found on the abaxial (lower) surface in some species. Each tentacle has a slender stalk that holds a round gland at the tip. The gland is enveloped with a droplet of sticky mucilage secretions. In sundews, the tentacles play a crucial role in the whole sequence of prey trapping and the subsequent digestion. In addition to these stalked glands (tentacles), numerous sessile (stalkless) glands are found on the leaf surface (and along the stalk of a tentacle in some species).

When prey, usually small insects and other arthropods, land on a sundew leaf, they immediately become mired in the sticky mucilage. The stimulated tentacles slowly move to bring the prey to the leaf center where digestion takes place. As the insect struggles to escape, nearby tentacles also begin to bend toward it. Although a typical tentacle movement is rather slow — taking a minute to a few hours to complete — the bending of tentacles undoubtedly enhances the sundew's chance of a successful catch. In many species the entire leaf gradually folds around the catch when a large prey is captured. Both the leaf-folding and tentacle-bending serve to bring more glands in contact with the prey.

The mucilaginous secretions appear to contain no lethal substances and the trapped insects typically die of exhaustion — or suffocation as mucilage covers their trachea during the struggle to escape.

It is unclear what attracts prey to the adhesive trap. Unlike pitcher plants that provide nectar, sundews do not offer any tangible reward to visiting insects. No aroma has been reported from sundew secretions. Insects, in general, are known to be attracted to shiny objects such as dews in the sun, and mucilaginous secretions of sundews may serve as a visual lure — or insects may be just looking for a place to land.

ABOVE: A large crane fly being overwhelmed by *Drosera anglica*, in a southern Oregon sub-alpine swamp, mid-June.

Digested remains on a leaf of *Drosera rotundifolia.* The original shape of the insect prey is almost obliterated. Note that the digestion is carried out by the glands of central tentacles — without the prey being dropped down on the leaf surface — because the stalked glands are capable of digestion and absorption. Sundews' sessile glands found on the leaf surface might be considered an evolutionary legacy, its function rendered obsolete, being replaced by the stalked glands. In early May, northern California.

DIGESTION

" Gentle strokes of tentacles paint the body of a prey with digestive juices.

The stalked glands of sundews are capable of producing digestive enzymes when a prey is captured. Upon physical as well as chemical stimulation, the secretions are switched from mucilaginous to digestive fluids. The sundews' tentacles are the only stalked glands (among all adhesive-trap carnivores) that have acquired this "dual" ability to perform digestive process as well — enzyme secretion and absorption — on top of adhesive prey capture.

In the pitfall traps, as we have seen, prey fall into a pre-formulated pot of digestive liquids. In contrast, the sundew's adhesive trap brings the digestive enzymes onto the prey: With a gentle stroke, flexing tentacles literally paint the insect body with digestive fluids, often covering the entire prey. This frugal application of metabolically expensive enzymes to targeted areas (by tropistic motion) contrasts sharply from the approach taken by the pitfall traps. Submersing the whole insect body requires such a large quantity of liquids that the majority of the pitfall traps rely heavily on external microorganisms to decompose the prey. Sundews produce all the necessary enzymes by themselves. Esterase, peroxidase, phosphatase and protease are among the enzymes found. The secretions also contain bactericides to prevent bacterial contamination during the digestive process.

As the digestion progresses, the insect body begins to dissolve. The stalked glands (and possibly sessile glands also) rapidly absorb the products of digestion. Research shows that, within as quickly as an hour, nutrients from the insect body begin to be carried to various parts of the plant.

TENTACLE STRUCTURE

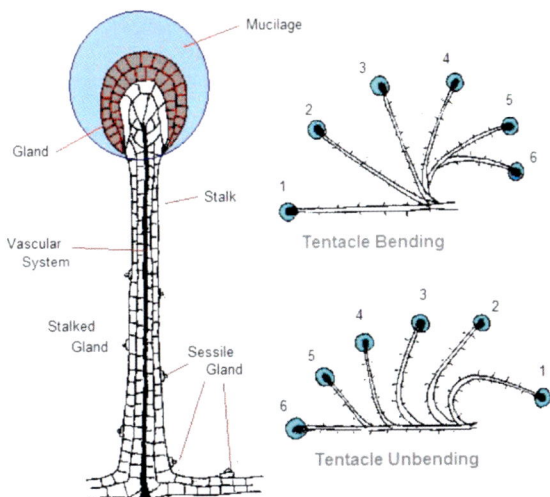

A close examination of the tentacle reveals that, unlike a typical plant hair which is of epidermal origin, the sundew tentacle has a far more complex structure exhibiting all elements of a real leaf itself. It is composed of a tall, tapering stalk of a multi-cellular structure, tipped with two layers of glandular cells. Mucilage is secreted from the glands through their cuticular gaps. A slender vascular strand exists in the stalk center that provides a gland-to-leaf linkage.

As described earlier, the tentacle is sensitive to physical stimulation and exhibits a bending motion. If the tip of a tentacle holding the gland is cut off, it no longer responds to direct stimulation (though the remaining stalk bends in response to indirect stimuli from nearby tentacles). This led to the traditionally held hypothesis that the sensor is located in the gland. It is now revealed that the epidermal cells located at the tip of the tentacle stalk (just below the two layers of glandular cells) are structurally homologous to the sensory cells found on the trigger hairs of the Venus flytrap.

Action Potential Upon physical stimulation to a tentacle — of sufficient strength for the receptor potential to rise beyond a certain threshold — a series of electrical pulses can be detected that rapidly travel along the tentacle stalk toward its base. These are action potentials. After a sufficient number of action potentials have passed to the motile region at the tentacle base, the tentacle starts to bend. The outermost marginal tentacles generate action potentials more readily and with greater amplitude and duration. The shortest tentacles in the leaf center require much stronger stimulation to raise their receptor potential to generate an action potential.

A crystal-clear droplet of mucilage covering the gland atop the tentacle tip of *Drosera filiformis* var. *tracyi*. The outer cells of the gland have a thin, perforated cuticle that allows passage of fluids. The epidermal cells on the upper tentacle stalk function as a physical sensor. Using *D. rotundifolia*, Darwin has shown that the tentacle responds to an object weighing less than 0.001 milligrams.

Tentacle Bending Mechanism The mechanism of tentacle bending is not fully understood. The bending is precipitated by a sudden drop of cellular pressure on one side of the stalk. The resultant imbalance of pressure between the opposite sides of the stalk causes the tentacle to bend. The pressure differential first occurs near the tentacle base, and gradually moves upward toward the tip.

After the absorption is completed, the tentacle returns to the original posture. It is the reverse process of bending, again due to pressure differentials. The unbending is a slow process usually taking a day or more. It is believed the unbending is a growth phenomenon seen as a result of restoring the cellular pressure. The same tentacle is measured to be greater in length by 10 percent or so after unbending. A given tentacle is capable of repeating the bending and unbending only a few times before it reaches its growth maturity.

Mucilage

Gland

Stalk

Vascular System

Stalked Gland

Sessile Gland

Tentacle Bending

Tentacle Unbending

Sundew Tentacle

An undisturbed leaf of a round-leaved sundew, *Drosera rotundifolia*, in May, northern California.

" **Tentacles respond differently to stimulation based on where they grow.**

Detailed studies have been conducted on the behavior of sundew tentacles by various authors, including Charles Darwin and F. E. Lloyd. Using the most widely known species, *Drosera rotundifolia*, tentacles can be divided into three groups based on their response characteristics to stimulation. Those tentacles found on the periphery of the leaf blade — usually the longest tentacles of all — are called the "marginal tentacles." Those on the leaf center are termed the "central tentacles" and are the shortest. Between these two groups of tentacles grow tentacles of an intermediate height, called the "outer tentacles." By and large, similar observations hold for the behavior of these three groups of tentacles in most sundew species, though some difference may be noted due to differing leaf morphology.

The marginal tentacle, when stimulated, always bends toward the leaf center. This type of bending motion toward a pre-determined direction, regardless of the origin of stimuli, is known as the *nastic* movement. When stimulated directly, the marginal tentacle does not transmit stimuli to other tentacles. This behavior distinguishes the marginal tentacles from the other two groups of tentacles. The marginal tentacle also responds to indirect stimulation from the outer and central tentacles.

The central tentacle, on the other hand, does not move when directly simulated, but does send out impulses to the neighboring tentacles in a matter of a few minutes. For indirect stimulation, the central tentacle bends toward the origin of the stimulus. This behavior of the tentacle — bending toward the direction which has a consistent correlation to the source of stimulation — is called the *tropistic* movement.

The outer tentacles, located in the middle zone between the marginal and central tentacles, exhibit somewhat of a combined characteristic of these neighboring tentacle groups. The outer tentacle responds to direct stimulation by exhibiting a rapid nastic motion, just like marginal tentacles. It will transmit an impulse to nearby tentacles as well.

For indirect stimulation, the behavior of both marginal and outer tentacles is more complex, typically showing an initial nastic movement followed then by a tropistic reaction, as if to correct the trajectory to reach the prey. The outer tentacles are known to be more responsive and agile to an indirect stimulus than the marginal tentacles.

The actual tentacle behavior may vary depending on where the stimulus has originated, on the different species, and on growing conditions. For instance, the higher the temperature, the more likely the tropistic reaction is to dominate.

In addition to these three tentacle groups, some ground rosette species, notably *Drosera glanduligera* and *D. burmannii* from Australia (and many others), produce long tentacles on the perimeter of the leaf blade. These tentacles do not produce any mucilage at the tip. Upon stimulation, the tentacles spring toward the center of the leaf blade — in a manner possibly tossing an unsuspecting passerby right onto the middle of the glandular leaf. The bending speed of these "snap" tentacles is much faster than that of regular tentacles and ranges from a lightning 1/30 of a second to a few seconds depending on the species.

The mucilage-tipped marginal tentacle of *Drosera intermedia*. Several sessile glands are seen along the tentacle stalk.

ABOVE FOUR: A leaf of *Drosera intermedia* showing tentacle-bending as well as leaf-folding in response to prey capture: 5 minutes, 10 minutes, 15 minutes, and 20 minutes after a small fly landed on the leaf.

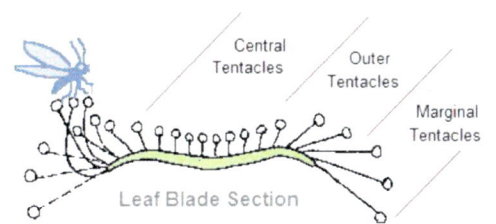

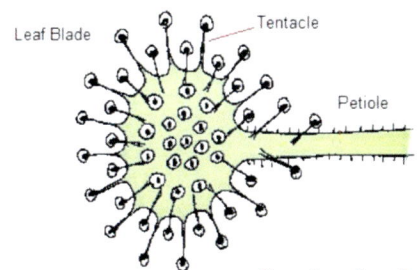

Central Tentacles

Outer Tentacles

Marginal Tentacles

Leaf Blade Section

Leaf Blade

Tentacle

Petiole

Sundew Leaf

Drosera rotundifolia

This sundew has a worldwide distribution throughout the Northern Hemisphere. In North America, the plant grows in the northern part of the U.S. and Canada. *Drosera rotundifolia* is commonly known as the round-leaved sundew. Charles Darwin used this sundew to conduct extensive experiments on the tentacle behavior.

The plant is perennial, forming a rosette of leaves, to 10 cm across, with a short stem. The leaf, that can reach 6 cm long in ideal growing conditions, is composed of a long petiole and a round leaf blade. The leaf blade, characteristically slightly wider than long, is covered with numerous tentacles, that assume a bright red coloration in plants growing in full sun.

The small, white flowers, 7 mm across, bloom in June to September in the U.S. habitats. The slender scape, to 15 cm, supports several flowers high above the prey trapping zone on the ground.

The plant typically grows in moist peat in bogs, sometimes in *Sphagnum* moss, always preferring an acidic environment. Plants tend to grow bigger and more vigorously in wetter conditions. One often finds the field covered with thousands of plants, creating large patches of glistening red carpets. In a typical natural habitat, mosquitoes, crane flies, moths, butterflies, and even dragonflies are seen successfully caught, sometimes by the cooperative effort of neighbors each lending a helping hand to overwhelm large prey.

In autumn, a hibernaculum is formed in the rosette center. This is a tightly packed firm bud that withstands cold winter.

ABOVE: A sighting of a damselfly being captured by the round-leaved sundews is not all that uncommon in the wild. In July, northern California.

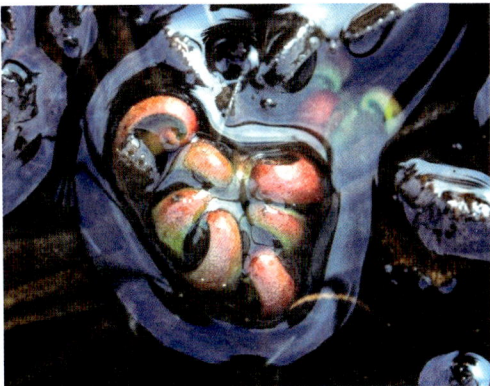

ABOVE: A winter hibernaculum of *Drosera rotundifolia*, in northern California, early December. RIGHT: Food-sharing. *D. rotundifolia*, in early July, northern California. OPPOSITE: For a small fly, any hope of escape wanes as the tentacles of the round-leaved sundew (*D. rotundifolia*) firmly secure the prey to the leaf center. In northern California, early May.

> **"Sensing the external stimuli, the sundew is *thinking* ...**

A small spider sits quietly in the center of a leaf blade of *Drosera rotundifolia*. Only a few tentacles have started to bend. Sensing the stimulus, the plant is "thinking" — analyzing the situation to decide the best course of action to take.

Like animals, plants can see (sense light), smell, and feel a physical touch. True — plants don't have eyes, noses, but their sensory apparatus is scattered over the entire plant. Plants do not possess a brain, as we animals do, but plants do respond to external stimuli in a highly coordinated manner — in spite of the absence of a centralized decision-making organ.

In the next hour or so, the tentacles will flex, the leaf blade may fold — in a most optimal way — to assimilate the prey.

In early May, northern California.

142

Drosera linearis

The plant is endemic to North America, recorded only in scattered localities in the northern U.S. and Canada. The areas around the northern Great Lakes region are historically noted for the abundant occurrence of this species. *Drosera linearis* is commonly called the linear-leaved sundew.

The plant is perennial, with a short stem, having erect, slender leaves, measuring 10 cm in length. The mature leaf has a petiole 3-5 cm long and a leaf blade 4-6 cm long. As the specific epithet (as well as the common name) suggests, the leaf is distinctly linear and the sides of the leaf blade remain parallel for a large part of the blade. This offers easy identification.

In the Great Lakes region, the plant flowers in late June to early July. The flower scape is generally short, 5-12 cm tall, barely exceeding the height of glandular leaves. Often a small number of flowers, 1-6 per scape, are produced. The white flower, 7 mm across, normally has five petals but six-petaled flowers are occasionally seen. The outer tip of the white petal may have a clear pink tint in some flowers.

The plants typically occur in marl fens, with the pH leaning toward alkaline. The soil substrate is sandy, mixed with marl, often with some surface water covering. In the Great Lakes region, accompanying carnivorous plants include *Drosera rotundifolia, D. anglica, Sarracenia purpurea, Utricularia cornuta, U. intermedia* and *U. minor.* In autumn, the plant forms a hibernaculum in the rosette center. This is a tightly packed firm bud that withstands the severe winter cold. As the warmth returns in mid-June the next season, the hibernaculum unfolds, resprouting new glandular leaves.

Found only in a narrow environmental niche, the highly restricted natural distribution of *D. linearis* is reflective of very particular habitat requirements of this species. The plant demands a pristine environment — no contamination or disturbance — and exhibits extreme intolerance for competition.

TOP: The five-petaled flower of *Drosera linearis.* The flower has five stamens and four two-lobed styles. (Some flowers have five, or even six, styles.) The tip of each petal is assuming a pinkish tint in this flower. The flower opens only for a few hours in bright sunlight. In mid-July, northern Michigan. BOTTOM: A bud with pink petal tips.

OPPOSITE: A big break for a small sundew. A damselfly struggling to free itself, in a marl fen in northern Michigan, mid-July.

ABOVE: Dewy leaves of *Drosera linearis* growing on the shore of Lake Huron where marl fens are formed. Note the relatively short flower stems typical of this species. BELOW LEFT: Distribution of *D. linearis* (red dots). BELOW RIGHT: One-year-old seedlings of *D. linearis* scattered around the mother plants. In northern Michigan, mid-July.

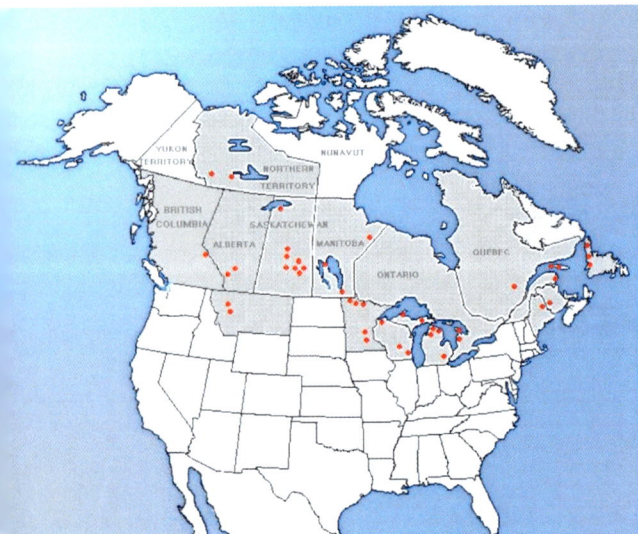

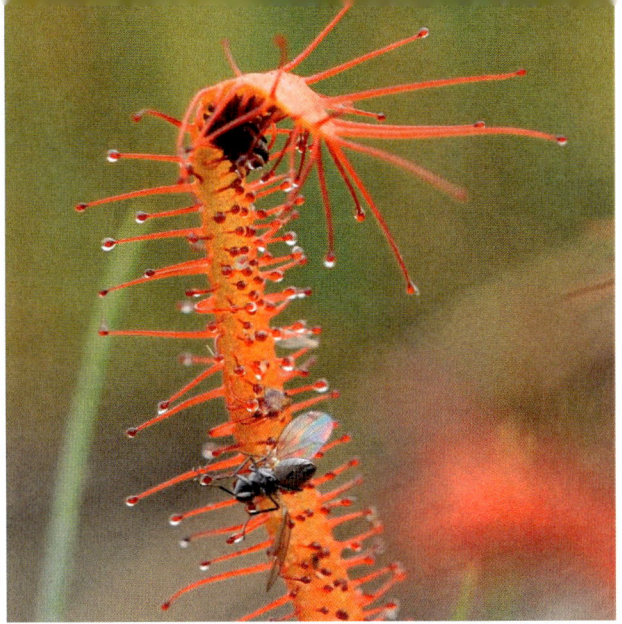

RIGHT: *Drosera linearis* capturing prey. In mid-July, northern Michigan. Note that the linear leaf of *D. linearis* is characterized by its sides being parallel.
BELOW: A typical marl fen habitat of *D. linearis* along the shore of Lake Huron, in mid-June, northern Michigan. Barren with sparse vegetation, the pristine habitat is home to many northern carnivorous plants including *D. rotundifolia, D. anglica, Utricularia cornuta, U. minor,* and *Sarracenia purpurea* ssp. *purpurea.*

A dense colony of linear-leaved sundews, *Drosera linearis*, in a marl fen along the shore of Lake Huron. The white flowers stay open only from 10 am to 2 pm. The plant blooms from June to mid-July in the Great Lakes region, and the seeds fully mature by August. Note the tiny seedlings on the soil. In mid-June, northern Michigan.

ABOVE: Glistening dewy leaves of *Drosera anglica* in a southern Oregon sub-alpine swamp, in early July. The plants prefer a very wet habitat, and the substrate is often covered with a thin layer of water. Unlike typical bogs in the Southeast, there is a constant flow of water. Three aquatic bladderworts, *Utricularia intermedia*, *U. minor* and *U. macrorhiza*, grow here though no other sundew species are found in this site. RIGHT: A white flower of *D. anglica*, in early July, southern Oregon. The flower has five petals, five sepals, five stamens, and three two-lobed styles.

150

Drosera anglica

The plant grows widely throughout the Northern Hemisphere, from North America to Europe and Japan. In the U.S. the plant grows in the Northeast and the Northwest, including the lower half of Alaska. *Drosera anglica* is commonly known as the English sundew, from the specific epithet — though it is unmistakably of North American origin.

The plant is perennial, forming a rosette of erect leaves, with a short stem. The leaves are narrow, measuring up to 15 cm long. A typical U.S. plant produces a leaf to 8 cm or so. The leaf consists of a petiole, 3-4 cm long, and a slender blade, 3-4 cm long.

The small, white flowers, 7 mm across, bloom in July to August in the U.S. habitats. The flower scape grows to 20 cm tall, but sometimes as short as the glandular leaves.

The plant is found in marl fens in the Great Lakes region where *D. linearis* grows. Elsewhere, *D. anglica* is found in northern bogs and swamps, sometimes in association with *Sphagnum* moss. The plant prefers a cool climate. In autumn, the plant forms a hibernaculum in the rosette center.

In 1955, C. E. Wood published a paper suggesting the hybrid origin of this species between *D. rotundifolia* and *D. linearis*. The natural hybrid *D. linearis* x *rotundifolia* is sterile, and can not bear viable seeds. However, in nature, a process called *amphiploidy* occurs spontaneously, doubling the chromosome number. This causes the plant to become fertile and establish itself as a new species. In fact, the speciation is on-going in multiple bogs in North America where *D. linearis* and *D. rotundifolia* grow sympatrically.

RIGHT: A colony of *Drosera anglica* in a sub-alpine swamp in southern Oregon, early July. The flowers bloom amid many glandular leaves on a relatively short flower stalk — though the bright, white flower petals do offer a sharp, visual contrast against the red trap leaves of the plants.

A majority of the prey for *Drosera anglica* are small gnats and other midgets, but a crane fly can occasionally fall victim to the long, adhesive leaves — which are fully capable of containing such an opportune situation. In early July, southern Oregon.

Unfolding of the winter hibernacula of *Drosera anglica* plants, in mid-June, southern Oregon. Snow has just melted a month earlier.

Sprouting new leaves of *Drosera anglica*, on a chilly, rainy day, in a sub-alpine swamp. In mid-June, southern Oregon.

Drosera intermedia

Drosera intermedia occurs in North and South America and Europe. In the U.S., the plant grows in the Atlantic coast and the Great Lakes region.

The plant is a perennial rosette, to 10 cm in diameter. Unique among all North American sundews, *D. intermedia* produces an elongated stem, if necessary, to keep up with the growth of surrounding vegetation. Sometimes a stem reaches 10 cm high. The long petiole, to 5 cm long, is round in cross section. The leaf blade is obovate, 5-7 mm wide and up to 20 mm long.

The small, white flowers, 7 mm across, bloom in June to August in the U.S. habitats. The flower scape grows to 15 cm tall and the seeds set by July.

The plant grows in *sphagnum* moss and in moist sandy peat, often in company with other carnivorous plants. In autumn, the plant forms a hibernaculum in the rosette center. A tropical form of *D. intermedia* plants is known that does not produce winter hibernacula and continues to grow all year round in a warm climate.

ABOVE: The spatulate leaf blade of *Drosera intermedia*. The long petiole is rod-like having a round cross-section. In mid-July, southern Michigan. RIGHT: The white flower has five petals, five sepals, five stamens, and three two-lobed styles. In mid-July, southern Michigan.

ABOVE: A dense cluster of *Drosera intermedia* plants, in a bog in southern Michigan, mid-July. The plants are growing in *Sphagnum* moss. Flowering appears slightly late this year in this habitat. BELOW LEFT: Ripe seed capsules of *D. intermedia* on a flower stalk. Many mature seeds are about to burst. In mid-July, southern Michigan. BELOW RIGHT: *D. intermedia* plants growing in wet, peaty soil. In May, in the Florida Panhandle.

Drosera capillaris

Occurring in North and South America and the Caribbean Islands, *Drosera capillaris* is the most commonly found sundew in the American Southeast. The plant occurs in the Atlantic coastal plain from Virginia all the way south to southeastern Texas, including the whole of the Florida peninsula.

The plant is perennial and forms a rosette 4-7 cm across. Larger plants tend to have raised leaves while smaller rosettes maintain prostrate leaves. The leaf consists of a slender petiole and a leaf blade, which is slightly longer than wide.

The small flowers, 1 cm across, bloom in May to August in the U.S. habitats. The flower scape grows to 15 cm and bears numerous flowers toward the top. The flower petals are usually white but pale pink is also common.

The plant is typically found on moist, sandy, peat soils, but can grow well in wetter areas. The plant survives in a dryer condition with a stunted rosette 2.0 cm across, and still manages to flower at that size. The plant assumes a bright red color in full sun. *D. capillaris* forms no winter hibernacula and endures the cold by protection from surrounding vegetation in the savanna grassland.

ABOVE: Flowering *Drosera capillaris* in Florida, in May. RIGHT: The five-petaled pink flower of *D. capillaries*. The flower color varies from white to pink. FAR RIGHT: Plants growing in a wet habitat in Florida, early May (top), and a plant growing on the moist sand surface, in Florida, early March (bottom).

The spatulate leaf of *Drosera capillaris.* Note the tiny colorless hairs on the leaf stalk.

Drosera brevifolia

Drosera brevifolia with a flower bud, in May, North Carolina.

The plant occurs in North and South America. In the U.S., the plant is found in the Atlantic coastal plain in more or less the same range as *Drosera capillaris* but appears to occur less frequently.

The small rosette rarely exceeds 3 cm across. Compared with *D. capillaris*, a wedge-shape leaf of *D. brevifolia* lacks a clear distinction of the leaf stem, and the petiole gradually expands into the leaf blade.

In spring, the plant produces a large flower, 1.5 cm across, relative to its small rosette size. The flower petals are pink to white. The short scape of this species is characteristically glandular.

The plants are found on moist sandy peat, often in company with other carnivorous plants such as *D. capillaris*. The flat posture of the glandular rosette often limits the size of prey and minute winged insects are a typical catch for this tiny sundew in nature. Plants growing in full sun assume an attractive reddish coloration. The plant does not form hibernacula. In North America, *D. brevifolia* may be annual in habit.

Drosera brevifolia is often found in the same general area where *D. capillaris* grows, though almost never in a very wet condition in which *D. capillaris* can occur. On this white, sandy peat soil, colonies of *D. brevifolia* and *D. capillaris* were found intermingled. In the sun, both sundews assume a very similar red coloration. The wedged leaf shape of *D. brevifolia* (RIGHT) that lacks a clear distinction between the leaf blade and leaf stem provides a reliable clue for identification. Note the long "snap" tentacles growing at the margin of a leaf blade that produce no mucilage. In the Florida Panhandle, early March.

Drosera brevifolia, growing on the forest floor, on a rainy day in May, North Carolina.

Drosera filiformis

Dwarfing all indigenous sundews in North America, *Drosera filiformis* is a tall, perennial plant with its erect, filiform leaves exceeding 30 cm in length. Commonly known as the thread-leaf sundew, the plants are found in abundance along roadside ditches and in savanna bogs.

Endemic to North America, *D. filiformis* has disjunct distribution ranges along the Atlantic coastal plain down into the Gulf Coast. Historically there are two expressions for this species each receiving an infraspecific varietal status: *D. filiformis* variety *filiformis* and *D. filiformis* variety *tracyi*.

Variety *filiformis* occurs in the northern range above North Carolina, and variety *tracyi*, along the Gulf Coast. Variety *tracyi* is larger and more robust-looking than the northern counterpart. Variety *filiformis* has a red tint in the gland while variety *tracyi* has totally colorless glands (though occasional, discernable reddish tint is present in some specimens).

Flowering starts in May and continues into June in the Gulf Coast. The large, conspicuous flower often exceeds 3 cm in diameter (for variety *tracyi*), having five petals, five stamens, and three deeply two-lobed styles. The pink flowers open in the morning, all facing in the southeasterly direction, and promptly close by noon.

The plant produces a hibernaculum in autumn. This is a large, tightly packed firm bud covered with dense white hairs for better protection from cold. In the warmth of spring, hibernacula sprout — in early March in Florida — and new glandular leaves unfold.

Drosera filiformis variety *tracyi*

TOP: Blooming *Drosera filiformis* var. *tracyi*, in southern Alabama, early May. All flowers are facing southeast in the morning sun. BOTTOM: The large, pink flower of *D. filiformis* var. *tracyi*. The lower three of the six stigmatic tips of this flower are colored yellow with pollen. The flowers open sequentially in the order from bottom to top along the flower stalk. Each flower opens only for one day. Note that an open flower is always positioned at the peak point of the hooked stalk. OPPOSITE: New dew-holding glandular leaves of *D. filiformis* var. *tracyi*, in southern Alabama, early May.

"Red" Variant In the midst of the variety *tracyi* distribution in the Florida Panhandle, small, isolated populations of plants quite similar to variety *filiformis* are known. The plants show pronounced deep-red coloration, providing marked contrast to the normal, eastern form. These deep-red variants are known to occur only in white sandy seeps around lime sinkholes. The color trait appears genetically fixed, and the striking red color is maintained in cultivation.

A propagule distribution by migratory birds is speculated to be the most likely method by which these disjunct *filiformis* populations had been established, but the question as to how they became so *red* is unanswered. This variant is now given formal variety status *floridana*.

Drosera filiformis
variety floridana

LEFT: A spectacular colony of *Drosera filiformis* var. *floridana* in the midst of a variety *tracyi* distribution in the Florida Panhandle. The disjunct populations of *D. filiformis* var. *floridana* are only found on wet, sandy, lime sinkholes by the lakes and ponds in the region. A stunning bright-red coloration of the glandular leaves easily distinguishes these populations from the normal variety *filiformis* in the East. In early May.

RIGHT TOP: *Drosera filiformis* var. *floridana* growing under the blazing sun, in a seemingly desert-like condition in the Florida Panhandle, early May. The soil is moist but very dry for sundews and many plants are stunted to mere 10 cm or so in height. RIGHT BOTTOM: A natural hybrid between *D. filiformis* var. *tracyi* and *D. filiformis* var. *floridana*. The hybrid exhibits robust leaf growth more typical of variety *tracyi* than a slightly slender-looking *floridana* variety. The color of the leaf tentacles is just about intermediate of the two. The plant as a whole appears attractive orange — contrasting sharply against almost colorless variety *tracyi* and bright-red variety *floridana*. This hybrid is not all that difficult to spot in the area where the two parents grow intermingled. In the central Florida Panhandle, early May.

Drosera filiformis variety *tracyi*

Late afternoon rays bring out a golden hue in the colony of thread-leaf sundews formed in a coastal savanna in the Florida Panhandle. *Drosera filiformis* var. *tracyi,* in late July.

A Venus flytrap with a flower bud, growing in peaty sand soil in native North Carolina, early May. OPPOSITE: A greenish specimen. With its massive modification of leaf morphology, the Venus flytrap must compete with other green plants for photosynthetic efficiency. In late July, North Carolina.

Venus Flytrap

GENUS *Dionaea*
FAMILY **Droseraceae**

Probably the most famous of all carnivorous plants because of its swift movement of trap leaves, the Venus flytrap is endemic to the Atlantic coastal plain of North America, where it is

highly localized to southern North Carolina and the adjacent northeastern South Carolina.

In a historical letter dated January 24, 1760, Scots-born Arthur Dobbs, then Governor of North Carolina, wrote to the English naturalist Peter Collinson, one of London's most influential horticulturists in the early eighteenth century, " … But the great wonder of the vegetable kingdom is a very curious unknown species of sensitive; it is a dwarf plant; the leaves are like a narrow segment of a sphere, consisting of two parts, like the cap of a spring purse, the concave part outward, each of which falls back with indented edges (like an iron spring fox trap); upon any thing touching the leaves, or falling between them, they instantly close like a spring trap, and confine any insect or anything that falls between them; it bears a white flower: to this surprising plant I have given the name of Fly Trap Sensitive."

This ignited a passion for this little plant from the New World. Many unsuccessful attempts followed in an effort to introduce living plants in Europe. In 1768, the Queen's botanist William Young finally succeeded in bringing live specimens to England and introduced them in Kew.

In the same year, a London merchant and naturalist, John Ellis, formally published the description of the plant in the *St. James's Chronicle*. His account was based on living material brought to England by Young. Ellis adopted the genus name *Dionaea* — from Dione, the mother of Aphrodite, the goddess of love and beauty, in Greek mythology — as suggested by his friend Daniel Solander, a Swedish botanist. To this Ellis added a specific epithet *muscipula* (mousetrap) with the English translation "*Venus's fly-trap*." Ellis was the first to suggest, definitively, the carnivorous habit of the plant. His enthusiasm is expressed in the letter sent to Carolus Linnaeus, " ... the plant, of which I now inclose you an exact figure… that nature may have some view toward *nourishment*, in forming the upper joint of the leaf like a *machine* to catch food: upon the middle of this lies the bait for the unhappy insect that becomes its prey."

Linnaeus, though he described the plant as the *miraculum naturae* (miracle of nature), dismissed Ellis' suggestion of carnivory — with unquestionable authority — and regarded such a notion as going "against the order of nature as willed by God." Linnaeus considered the movement merely an irritability as seen in *Mimosa*, and believed that a trapped insect was released upon reopening of the lobes.

The genus *Dionaea* is monotypic, consisting of only one species, *Dionaea muscipula*. The genus is placed in the sundew family Droseraceae.

DESCRIPTION

> " One of the
> most wonderful
> in the world.
>
> — Charles Darwin

Traps with deep-red inner lobes, creating a powerful visual allure for unsuspecting insects. *Dionaea muscipula* lies in wait in native North Carolina, late July.

The plant is a perennial herb forming a rosette. The leaves emerge from a short rhizome (underground stem) with black, fibrous roots. The white, thickened base of a leaf is buried underground. The leaves, which can grow to 10 cm long or more, are of two parts: a flat petiole (leaf stem) and a leaf blade modified into a trap. The leaves in the early spring are short and tend to have a broader petiole lying flat on the ground. Further into the season, the leaves often become semi-erect (possibly in response to increased winged insect population in summer) and the petiole becomes narrower (possibly due to increased photoperiod). This is only a general rule, and some strains consistently pro-duce a flat rosette of wide, decumbent petioles while some maintain narrow, raised leaves all year round.

The trap portion consists of two semicircular lobes united along the midrib. Around the margin of each lobe grow 15-20 stiff spines, or marginal teeth. Along the inner edge of the lobe just below the spines runs a narrow band of nectar glands which secrete sugary substance to attract potential prey. Much of the inner lobe below is crowded with numerous digestive glands.

The nectar and digestive glands are identical in structure, though distinct in function, with each gland consisting of about 32 glandular cells. The digestive glands are slightly larger in size and are far more numerous than the nectar glands. Both are sessile (stalkless) glands. Nectar glands are often colorless while digestive glands are typically heavily pigmented which color the trap interior with a deep-red hue.

The inner surface of each lobe has three — sometimes four or more — fine bristles located in a triangular pattern. These are trigger hairs, which are sensitive to physical stimulation and, when properly stimulated, initiate rapid trap closure. Studies have revealed that the trigger hair has sensory cells embedded at an indentation near the base, where bending strains are most pronounced when the hair is disturbed.

Note that there is a narrow, gland-free strip on each lobe that lies between the nectar-gland and digestive-gland regions. After prey capture, the two lobes touch each other along this strip and form a tightly sealed trap cavity before digestion starts.

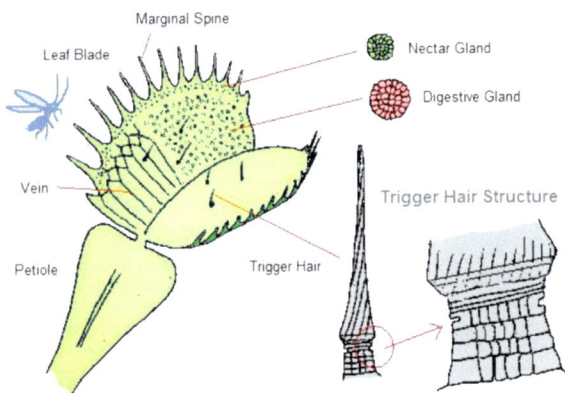

Venus Flytrap Leaf

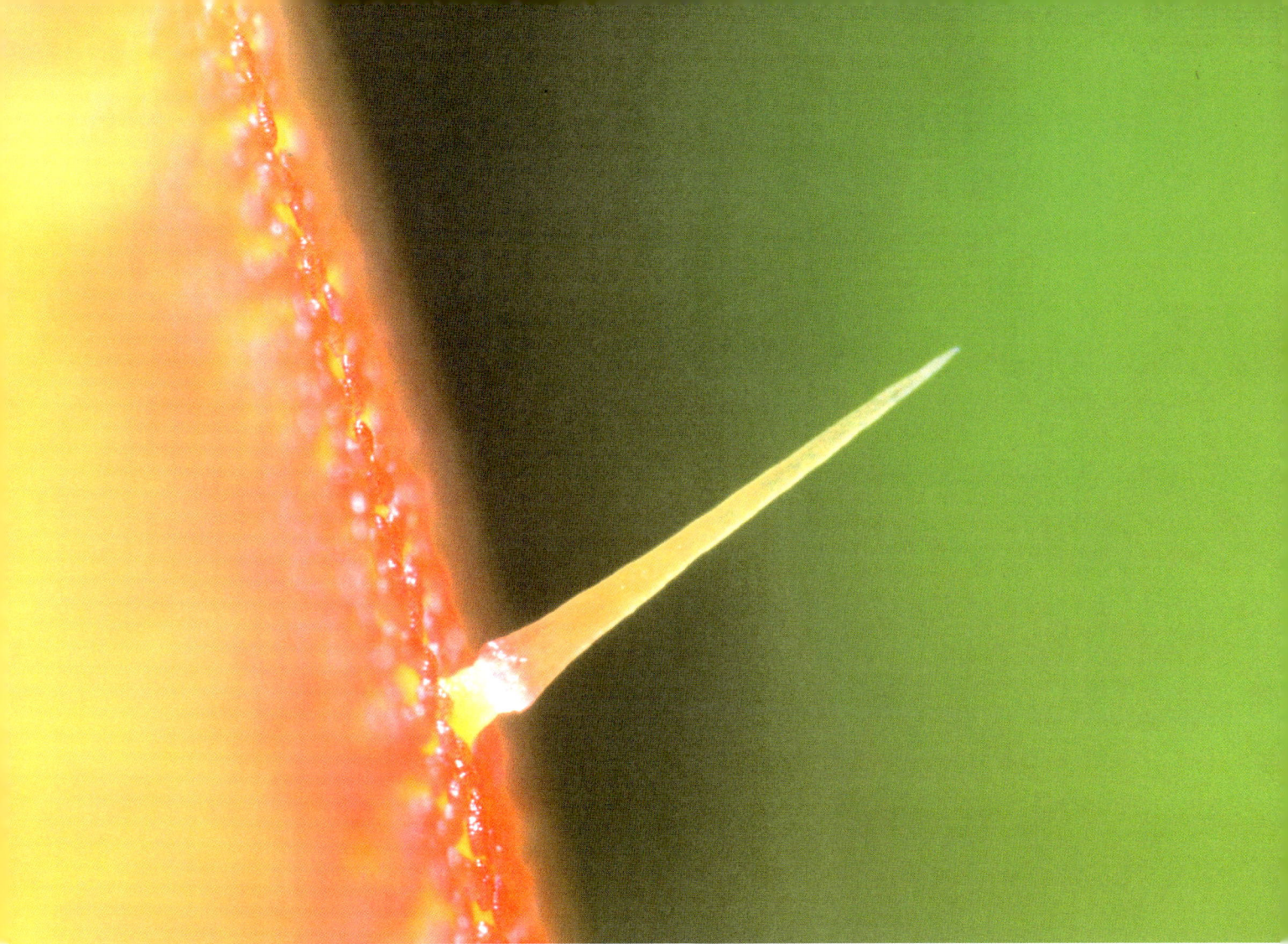

ABOVE: The trigger hair of the Venus flytrap. Frail and benign-looking — and almost unnoticed by foraging insects — these tiny hairs on the lobe surface are a highly sensitive triggering apparatus. When disturbed by a potential prey, sensory cells embedded near the hair base generate an action potential, commencing a deadly trap closure sequence.
BELOW LEFT: The inner surface of a Venus flytrap's lobe showing numerous digestive glands. In this leaf the greenish venation is clearly visible where the digestive glands end. Farther above towards the marginal spines are scattered nectar glands that secrete sugary liquid as a ruse.

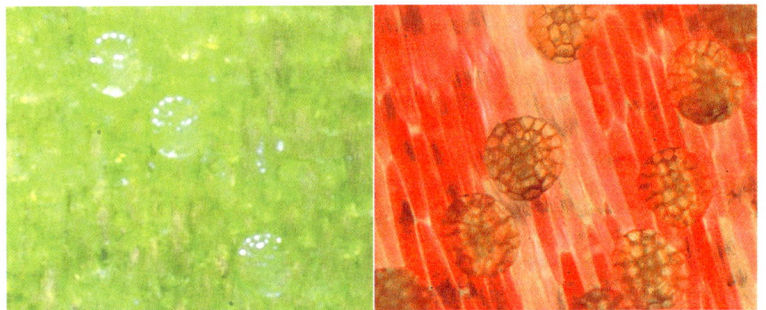

ABOVE LEFT: Colorless nectar glands scattered along the lobe margin. ABOVE RIGHT: Numerous digestive glands are often tinted with a bright red color. Each gland comprises multiple glandular cells. Note that the epidermal cells on the lobe surface are longer in shape in the direction perpendicular to the midrib of the trap.

In a colorful botanical history of the introduction of the Venus flytrap into Europe, mischievous eighteenth-century botanists — John Bartram, a distinguished plant collector of Philadelphia, being a central conspirator — saw an apparition of female anatomy in the trap appearance, and in the correspondence to one another, the plant was referred to by a curious name "tipitiwitchet," which Ellis surmised to be a local Indian term, though the linguistic origin seems to point to Elizabethan vernacular. Aphrodite's mousetrap, *Dionaea muscipula*, in humble admiration of the goddess of love and pleasure. In July, North Carolina.

INFLORESCENCE

> " **Innocent grace emanating from elegant, white flowers masks the true, murderous nature of the plant.**

LEFT: The white flowers of *Dionaea muscipula* bloom on a tall stalk. Each flower stays open for a few days. BELOW: On the first day, the pollen-receiving stigma in the middle of the flower appears fist-like and does not become receptive immediately.

On the second day, the stigma opens to receive pollen. Note the hairy fimbriate appearance of the stigmatic tip. Ellis, in naming the specific epithet *muscipula*, did not realize Solander had intended the name *crinita*, referring to this fimbriate stigma of the flower.

In early spring, a flower bud appears in the rosette center. In a month or so, this develops into a tall stalk bearing many white flowers at the top. The stalk often reaches a height of 20-30 cm, separating pollinators away from the trap. Flowering occurs in May through June in native North Carolina.

Graceful white flowers set high on a flower stalk offer no hint of the true, vicious nature of this carnivore — yielding a diabolically stark contrast to the bloody massacre on the ground.

The actinomorphic flower (radially symmetric) has five petals and five sepals. The white petals are shot with greenish veins. Fifteen or so stamens surround the ovary.

Each flower typically remains open for a few days. The flower is protandrous and the stigma does not become receptive until the second day. This would encourage cross-pollination though viable seeds are produced by selfing. After fertilization, the ovary swells and numerous black seeds fully mature by late July. As the seed capsule dries, the pressure from capsule shrinkage bursts the slippery, pear-shaped seeds around the field. The seeds are large for the carnivorous plant standard, measuring 1 mm in length. In nature, the seeds germinate shortly after dispersal and tiny seedlings have a few months to prepare for the onset of winter. In spite of the relatively small size of the plant, it takes three to four years for the Venus flytrap seedling to reach a flowering age.

LEFT: Birth of a Venus flytrap. Germination commences as the seed absorbs water and swells. Embryonic roots anchor the seedling in the soil. RIGHT: One-week-old seedlings with their twin cotyledons just emerging.

One-month-old Venus flytrap baby with the first trap. The black, pear-shaped seed of the Venus flytrap is 1mm long. When the seed germinates, a baby trap leaf emerges right after twin cotyledons (seed leaves). The trap is equipped with small trigger hairs as well as some digestive and alluring glands, and it is fully functional to trap a bug matching its size.

A point of no return — A fly is seen at the moment of a deadly snap. For the victim inside, the whole world collapses upon it with overwhelming force and speed. The marginal spines intermesh, effectively entrapping a prey of a *worthy* size. Trap leaves close swiftly and with such calmness, even the most agile of flyers finds no time to jump out of harm's way.

The trap of the Venus flytrap is arguably the most advanced among all carnivorous plants. This has not come without a price. A large portion of the plant's biomass is diverted to the construction of a highly sophisticated trapping apparatus. The dynamically active trap also exacts a high operating cost during its life. To mitigate the energy consumption, a series of subtle but clever strategies are in place in the trap operation, as we shall see.

Attraction The plant catches prey quite effectively in the wild as well as in cultivation. Typically, arthropods — insects and other small bugs, like spiders — are a typical diet for the plants in nature. Many catches are not an accidental passer-by. They are deliberately guided to the trap by attractants.

In addition to the brilliant color of the trap, sweet nectar is surely an important ruse for alluring prey. When the plant is in a

" Multiple stimulations requested: the fatal, *second* caress.

vigorous growing condition, a narrow band of wet surface is visible along the inner margin of the trap lobes. This is due to secretions from nectaries (nectar glands). Anyone who has observed an insect licking the nectar will be convinced that there is something magical about the secretions. Research has found some narcotic substances in the exudation that intoxicate the prey. The nectar band has also been shown to be UV-absorbing, making the band look distinctively dark in insect vision.

Strategic Trigger Hair Placement With its mouth working on the nectar just below the marginal spines, an insect large enough in size to brush against some of the trigger hairs ends up tripping the deadly trap. The trigger hairs are so situated in relation to the nectar band that, given a mature trap of 20-30 mm in length, an insect measuring 5 mm or less is not likely to touch the trigger hairs. Considering the amount of energy needed to close the trap, this preliminary prey screening provides effective cost management for a tiny plant. Lloyd cites a report showing that, of fifty closed traps examined, only one contained the catch less than 5 mm in length, and only seven less than 6 mm. All the other catches were 10-30 mm long.

Multiple-Stimulation Requirement Many experiments have been conducted on the snap-trap behavior and it is widely known that a trap requires two separate touches to close. In a normal condition, it is necessary to stimulate two different trigger hairs, or stimulate the same hair twice, within a 20-seconds or so interval. If two stimuli are given this way (minimum one second apart), a very swift closure — well under half a second — immediately follows. Without eyes to monitor the insect's whereabouts in the trap, this is a plant's clever way of assuring that a potential prey has been well situated in the trap center before triggering the action. This may also reduce the chance of wasteful closure by wind-blown debris. The longer the interval between the two stimuli the less rapid the closure becomes. If the second brush comes after a half minute or

longer, the trap movement itself becomes sluggish and many stimuli may be required to complete the closure.

On the other hand, on a sizzling summer day, a trap may close by a single touch — an accidental firing. It is also shown that the stimulation to a lobe part other than trigger hairs, particularly the area near the hairs, sometimes induces trap closure.

> **" All visitors, please line up for measurements.**

If a trigger hair is deliberately stimulated by an inanimate object — or by a curious finger — the trap also snaps shut. In this case, the trap opens the next day, since no nutritious object is found inside the trap. A trap can be "fooled" to close ten or so times before it reaches its growth maturity and becomes insensitive to stimulation.

Two Closure Phases There are two phases in the trap closure. The initial phase is characterized by a rapid snapping of the trap lobes that brings the marginal spines on both lobes together, close enough to interlace them. This will effectively entrap the prey of a sufficient size while leaving some gaps between the spines for a smaller prey to sneak out — the second screening of prey, if you will — before proceeding to a metabolically intense digestive process.

LEFT: Chitinous remains of a victim left in the trap after the absorption is completed. In July, North Carolina. One report on captured prey in North Carolina has shown an overwhelmingly large proportion of the catch being crawling or ground insects and other arthropods (notably spiders and ants), causing some to question the appropriateness of the name "flytrap," while another report indicated a much higher capture rate of winged insects. RIGHT: A spider on a trap, having passed the "size requirement" test for food-worthiness with flying colors.

DIGESTION

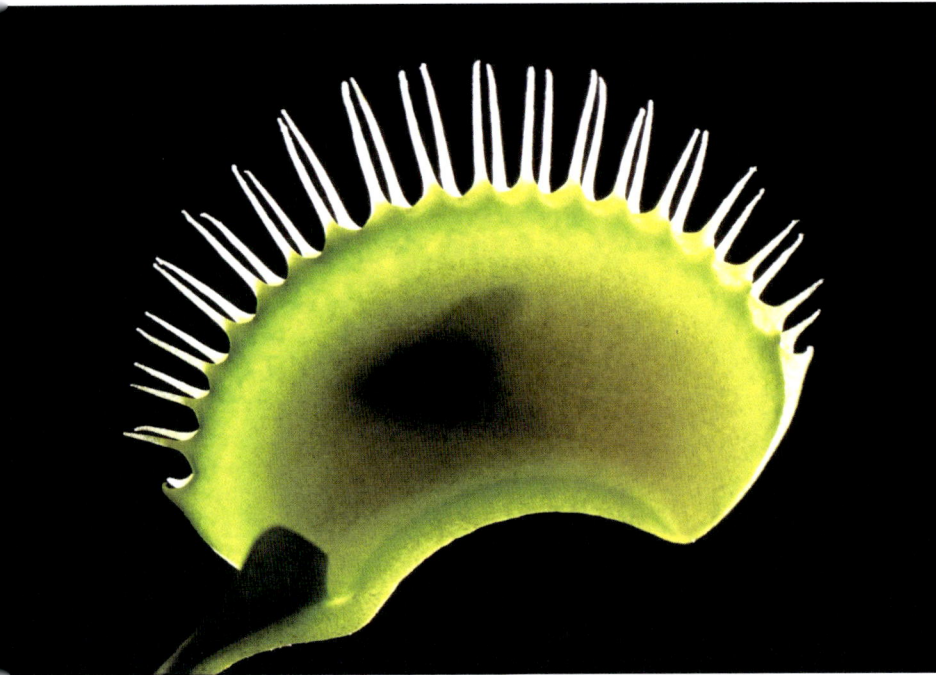

" The Venus flytrap does not rely on external organisms to complete the digestive process.

The initial rapid closure is followed by what is termed the narrowing phase. This is pursued only if the trap continues to receive physical stimulation, as would inevitably be the case if a live insect is captured. The narrowing phase is a slower process which brings the margin of the two lobes tightly together. Often the pressure exerted in this phase is strong enough to crush a soft-bodied insect prey.

As the lobes are sealed, the glands start to secrete digestive fluids into the closed trap chamber. The prey, if still alive, will most probably be suffocated in the liquid. The secretions, which are known to be initiated only by proper chemical stimulation, typically begin several hours to a day after prey capture, and reach the maximum quantity in 4-7 days. Uric acid, in particular, the major constituent of insect excrete, was found to induce a large volume of enzyme secretions. The enzymes detected in the secretions include protease, phosphatase and amylase.

The insect body begins to dissolve in the liquid. The digestive process, not unlike that in an animal's stomach, lasts for a week or so, depending on the size of the prey. The secretions from the glands contain antiseptic substances that effectively prevent bacterial growth during the entire digestive process. The digested material is absorbed through the glands as well as the inner surface of the trap.

The digestion and absorption accomplished, the now dry trap reopens slowly, revealing the undigested exoskeleton of the prey. The wind and rain will clean the lobes of chitinous remains, and the trap is ready again for the next meal to walk by. Too large a catch often results in the damage of a trap, as the exposed portion of the animal begins to decay, causing the entire leaf blade to blacken and putrefy before the digestion is complete — an upset stomach, as it were.

A single leaf is capable of repeating this digestive cycle only a few times during its life. After that the leaf becomes nonresponsive to stimulation and remains as a photosynthetic organ until it dies away. New trap leaves continuously sprout from the rosette center during the growing season from spring through summer.

TOP: A back-lit trap reveals the shadow of a fly under digestion. OPPOSITE: Bear traps left on the floor — Venus flytraps on the forest margin, seemingly overgrown by surrounding vegetation. In the absence of prescribed burns, a prime carnivorous plant habitat undergoes transformation into an advance forest, forever annihilating a niche environment these unique plant species are adapted to. In July, North Carolina.

" **Over 250 years after its discovery, many unanswered questions remain on the trap closing mechanism.**

A virgin trap.

CLOSING MECHANISM

S tudies have been conducted to better understand the trap closing mechanism of the Venus flytrap. To this day, many unanswered questions remain.

Upon stimulation to a trigger hair, the sensory cells embedded near the base of the hair generate an action potential that propagates swiftly (0.2 sec.) over the entire trap lobes.

Memory The action potential activates the gene that encodes the "calcium channel" protein, causing a rapid inflow of Ca^{++} ions into the cell's interior along the propagation path. This elevates the Ca^{++} concentration of the entire trap to a high level. This may account for the "memory" mechanism of the Venus flytrap to remember the first stimulus. When a trigger hair is touched again (causing the second wave of action potential) the leaf's Ca^{++} level is pushed higher to a set threshold and a trap triggering action is initiated. The Ca^{++} concentration (memory) starts to decline steadily right after the maximum level of the first touch and a second stimulus must follow in a timely manner — within 20~30 sec. — to bring the Ca^{++} concentration to the triggering threshold. Note also that the second action potential must not arrive too soon while the first calcium channel is still open (1 sec.), in which case the second touch is ignored. This explains why the trap does not close if two stimuli are given to trigger hairs in rapid succession.

Expansion It is not completely understood what really triggers trap closure, and exactly what happens during the time of closure. Unbeknown to the insect having caused the fatal, second stimulus, a silent alarm ripples through the trap floor, setting in motion a deadly trap closure sequence.

A moment prior to the trap closure, the entire lobe interior (mesophyll between the upper and lower epidermis) seems to expand precipitously. The lobe expansion occurs only in the direction perpendicular to the midrib. Meanwhile only the lower (abaxial) epidermal walls become "loose" and extensible. This creates pressure for the lobes to warp inward.

Marginal spines on both lobes are lined up precisely so they mesh when the trap closes. Spacing of the "teeth" defines the minimum prey size worthy of swallowing.

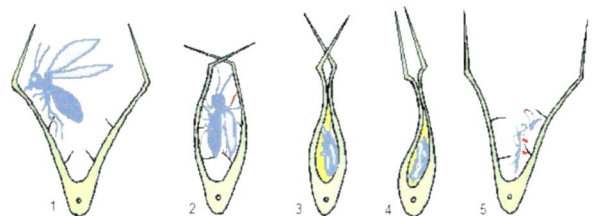

Venus Flytrap Leaf Closure Sequence

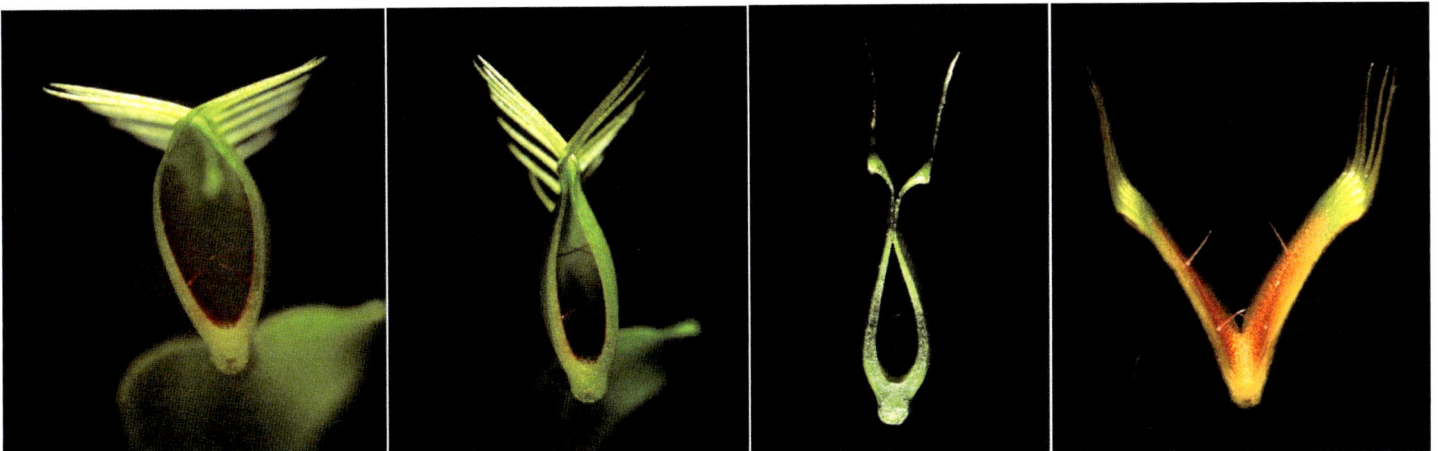

LEFT TO RIGHT: The initial swift closure is followed by a slow, narrowing phase that brings the two lobes tightly together. As the narrowing continues, the trap cavity becomes increasingly small, and the pressure is likely to crush soft-bodied insects to death. After the digestion is over, the trap opens slowly. Note the more opened angle of the marginal spines, compared to the initial posture of a virgin trap.

TOP: An open trap, with the convex curvature of the upper surface. BOTTOM: Upon trap closure, the lobes flip, reversing the surface curvature from convex to concave.

If one lobe is cut off from the trap, the remaining lobe curls more than 180 degrees.

" After the second jostle to a trigger hair, the strain rapidly builds in the lobes ...

A Sudden Flip When the trap is open, the inner lobe surface is slightly convex. Because of this initial surface curvature (double-curved surface), the pressure due to the lobe expansion upon triggering does not immediately produce an inward warping of the lobes, and the trap remains open. The lobes resist the build-up of this mechanical tension for a short period, say, 1/3 of a second.

The increasing strain now reaches an unsustainable level and, yielding to the stress, the lobes flip, precipitating a sudden change in curvature from convex to concave (as seen from above). This results in a swift closing motion of the two lobes. In a typical, healthy trap, this "snap-through" buckling does contribute to some extent to the apparent increase in the lobe closure speed.

It should be noted, however, that some traps may not have a sufficient initial lobe curvature to induce appreciable flipping of the lobes. Even in these traps, the closure can occur reasonably fast, say, within half a second, due to the rapid expansion of the lobes. In this case, the trap closing motion will start immediately upon the second stimulus to the trigger hair, without waiting for the flip to occur.

In general, the lower epidermis near the lobe center is stretched by as much as 7 % in the direction normal to the midrib.

Narrowing Phase The lobe expansion, with only the lower epidermis stretching, continues into the narrowing phase, exerting the pressure to close the two lobes more tightly. The pressure is so strong that if one lobe is removed (cut off) from the trap, the remaining lobe curls more than 180 degrees. This pressure seals the trap cavity before the commencement of the digestive process. To ensure a tight seal, the contact surface area of each lobe — a narrow strip just below the nectar glands — is void of any gland.

Reopening After the absorption of digested products, the trap reopens slowly, taking a day or so. Reopening appears opposite to closing, and only the upper (adaxial) epidermis becomes extensible, and stretches as much as 7 % in the direction normal to the midrib. Reopening is due to cell growth. Reopened lobes have been measured to be several percent larger on average than the original size prior to trap triggering.

OPPOSITE: Multiple-exposure photography reveals a swift change of the angle of marginal spines during the initial rapid snapping, effectively preventing prey escape. INSERTS: The fly capturing sequence. The last picture shows the same trap two days after the catch.

" … And then, a sudden and deadly flip follows.

NATURAL HABITAT

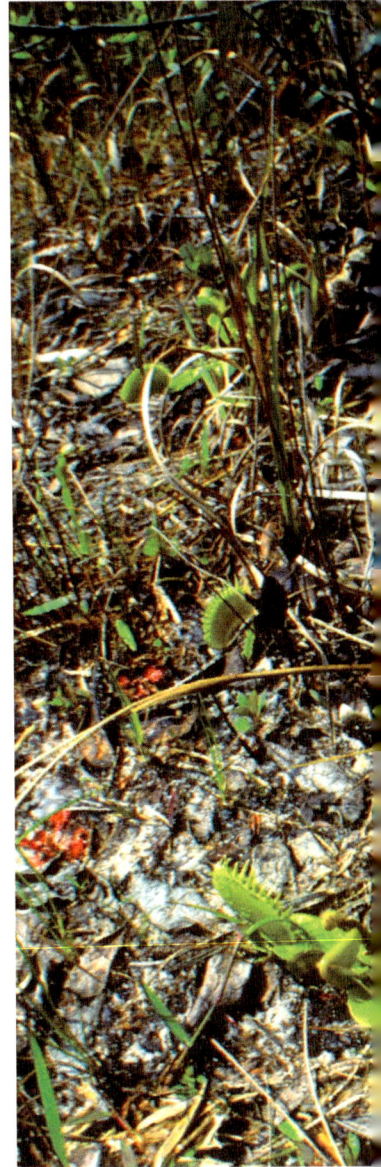

The natural distribution of the Venus flytrap is confined to the coastal plain of North and South Carolina — within a 100 km radius around the city of Wilmington on the North Carolina coast — and the range is rapidly shrinking due to habitat destruction. It goes without saying that collection from the wild is utterly unconscionable and is against the law with a stiff penalty. Luckily, tissue-cultured specimens are readily available for curious minds.

The plants are typically found on moist soil in the open pine forest, sometimes on a mat of *Sphagnum* moss, sometimes on a white sand surface. Accompanying carnivorous plants found in the general vicinity include sundews, butterworts, terrestrial bladderworts, and some pitcher plants.

Dionaea is a sun worshipper and grows best in an open, sunny locality. To maintain the healthy growth, periodic wildfires of the habitat are essential to clear the encroaching vegetation that can easily rob the plant of vital sunlight. *Dionaea* is intolerant of competition with other plant species, and it does not take long for the population to be wiped out. The Venus flytrap enjoys acidic, nutrient-poor soil where competition is low.

" Native populations are rapidly declining due to habitat destruction.

The climate of the habitat is temperate; summer is hot and humid while winter may experience light freezing. The area may be inundated with flood water from occasional storms. The plant seems to withstand prolonged flooding quite comfortably. Some sightings of aquatic catches are reported on such occasions.

There are places in the United States (and elsewhere) where Venus flytraps were transplanted some time ago and are well established since. Florida has a few such populations. As for the natural range, it is puzzling why *Dionaea* failed to extend its distribution farther south, considering there are vast stretches of seemingly prospective habitats along the Atlantic coast all the way down to the Gulf Coast.

One of Florida populations transplanted many years ago. The colony is well established since with hundreds of healthy plants and many seedlings. In the central Florida Panhandle, early May.

An Evolutionary Dead End Many botanists consider *Dionaea* to be the most advanced of all carnivorous plants in terms of its remarkable adaptation. This evolutionary triumph did not guarantee the Venus flytrap a secure future. The highly specialized trap mechanism came at a high cost of meager ecological flexibility. Sundews (the genus *Drosera* in the same family Droseraceae) are faring superbly better in their diversity and worldwide presence. Given the same environmental stress — man-inflicted or otherwise — the Venus flytrap seems to be losing out in a test of survival of the fittest.

In a twist of nature, human curiosity offered a helping hand to this "miracle of nature." Today, Venus flytraps are omnipresent throughout the world. The critical underpinnings of any organism rest on the sheer number and future security. By this measure, one might argue, the Venus flytrap may have succeeded. But *in vitro* cloning in nurseries is hardly a *natural* process, unlike help from bees and other pollinators. Because we humans are *above* nature — or, are we?

Young Venus flytrap plants growing in peaty sand, in native North Carolina. Also seen are red rosettes of *Drosera capillaris*. In early May.

185

Butterworts

GENUS *Pinguicula*
FAMILY **Lentibulariaceae**

Butterworts use a flypaper, or adhesive, trap to capture small prey. The surface of a leaf is covered with numerous fine hairs which hold a sticky glue. The leaf feels greasy to the touch

---

because of the secretions. This gave the plants their common name. The genus name *Pinguicula* is derived from the Latin word *pinguis*, meaning fat.

Butterworts typically grow in grass-covered sandy soil in savannas, on mossy rocks, and in other similar, moist conditions.

There are some 90 species of butterworts in the genus occurring worldwide, with Mexico and Central America harboring more than fifty species. Central America, where the center of diversity is seen today, is speculated to be where the genus originated. By and large, each species of *Pinguicula* occupies a relatively narrow range. The secondary diversification is suspected to have taken place in Mediterranean Europe, leading to the circumboreal distribution of the *younger* species, *P. vulgaris* and *P. villosa,* that occur in many parts of Europe and in North America. *P. macroceras* (closely related to *P. vulgaris*) occurs in boreal North America, Japan and Russia. *P. alpina* also has a wide distribution, found in much of Europe, Russia and Asia. The southeastern part of the United States contains six species, five of which are endemic to the region.

Apart from the carnivorous nature of the butterwort plants, people in Scandinavian countries have mixed the leaf extracts with fresh milk to make it curdle. Leaf extracts are also said to have a healing effect on wounds, and Europeans have applied butterwort leaves to the sores of cattle.

TOP: The purple flower of *Pinguicula caerulea* in Georgia, late April.
OPPOSITE: *P. macroceras* in flower. In southwestern Oregon, mid-April.

DESCRIPTION

Butterworts are a compact rosette of thin leaves attached to a short stem, with white, succulent roots. The leaves typically lie flat on the ground. The upper surface of the leaf (and peduncles in some) is covered with glandular hairs tipped with mucilage.

Many butterwort rosettes have close resemblance to one another and it is often difficult to identify the species when the plant is not in flower. The leaf is yellowish-green in most species, but can assume a distinctive reddish tint in some. The leaves are generally flat but may be slender and erect in some species. The rosette diameter varies anywhere from 2 cm to 20 cm depending on the species.

The majority of butterworts are perennial. Many northern species form winter hibernacula during the cold months. A hibernaculum is a small, tightly packed, bud-like structure that tolerates severe cold and desiccation. (This becomes entirely rootless and, therefore, may be displaced by water and wind.) The temperate species of the southeastern United States maintain glandular carnivorous leaves year-round without forming hibernacula. Many Latin American species produce scale-like succulent leaves during the dry season. These winter rosettes, smaller than normal carnivorous leaves, are often hairy, but no glandular secretions are present.

Two species (*Pinguicula gigantea* and *P. longifolia* subspecies *longifolia*) are known to have adhesive hairs on the lower surface of the leaves as well.

BELOW: Light-green rosettes (6 cm across) of *Pinguicula vulgaris* in northern Michigan, along Lake Huron. In mid-July. OPPOSITE: The yellow-flowered butterwort, *P. lutea*, bearing a large, bright yellow corolla on a tall scape. In Florida, early March.

189

INFLORESCENCE

Butterworts are known for a vibrant display of colorful flowers which are borne singly in a pendulous position at the tip of a slender, often glandular, pubescent scape (*Pinguicula ramosa* being the sole exception having branched flower scapes). The calyx has five sepals. Early spring is the flowering season for many U.S. species. The zygomorphic (bilaterally symmetric) flower has a sympetalous (united petals) corolla forming a cylindrical tube, that divides into five petals made of a two-lobed upper lip and a three-lobed lower lip.

The corolla tube terminates in a slender spur on the back. A spur is a nectar container, a structure commonly seen in insect-pollinated flowers. On the lower entrance of the corolla tube protrudes a hairy structure called a palate. The palate hairs are edible body tissues, decoys that divert the attention of non-pollinators away from precious pollen. The microscopic structure of the palate hairs (and other hairs around the tube entrance) is characteristic to each species.

The flower opens nodding, and at the floral base (on the upper side of the inner corolla tube) sits a glandular, spherical ovary surrounded by two stamens. From the ovary tip grows a short style ending in a two-lobed stigma. The wide, lower lobe of the stigma completely covers the anthers, exposing the pollen-receptive stigma surface for an incoming pollinator to deposit pollen. As the pollinator exits the flower, the stigmatic lobe flips, preventing the flower's own pollen to be deposited to the stigma again — a mechanism to discourage self-pollination.

After fertilization, the scape grows further as it straightens itself. The seeds mature in a month or so. The seed capsule contains hundreds of powder-like seeds. The seeds have a distinctive surface pattern that provides a taxonomic character.

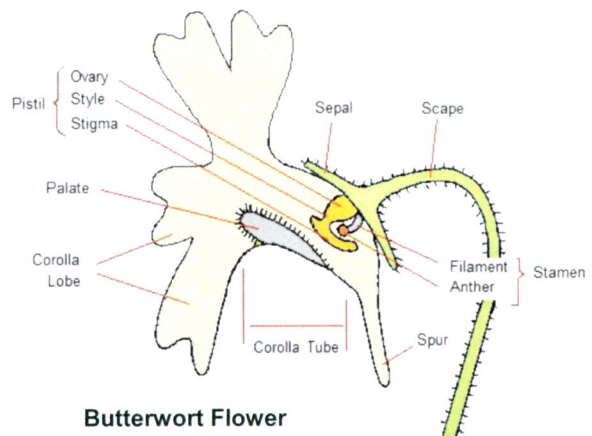

Butterwort Flower

"**Many butterworts resemble one another and the distinction is often difficult without flowers.**

Pinguicula ionantha flower with a strongly exerted yellow palate. In Florida, in early March.

RIGHT: A *Pinguicula primuli-flora* flower with the lower corolla lip removed to show its reproductive anatomy. The two arched stamens surround a round ovary. The thin stigmatic lobe completely covers the anthers, with the gooey receptive surface exposed. FAR RIGHT: A mature seed pot of *P. vulgaris.*

OPPOSITE: A blooming colony of *Pinguicula ionantha* plants in Florida, early March.

191

Pinguicula vulgaris growing on the moss-covered rock surface on a cobble beach along the shore of Lake Huron. Spring has arrived a bit earlier this year, and many seed capsules are already forming at the tip of the elongated scapes. In northern Lower Michigan, mid -June.

ADHESIVE TRAP

Most of the prey butterworts are capable of trapping are small winged insects such as gnats and midges. As in sundews, there is no food rewards for visitors. What attracts prey to the butterwort trap is not clear. Perhaps the glistening look of the leaf surface provides allure. A faint smell of the leaf may be an attractant.

Tiny hairs covering the leaf surface are composed of a gland supported on a slender, single-celled stalk. Each gland comprises 16 radiating glandular cells. These stalked glands produce mucilaginous secretions for this adhesive trap.

> " There is no take-home reward for the visitors in the butterworts' adhesive trap.

The mucilage oozes out through cuticular gaps in the glands and forms a clear, spherical droplet at the tip of each stalk.

Scattered over the leaf surface is another kind of gland with a similar structure. These are sessile (stalkless) glands consisting of 8 radiating glandular cells, and are almost buried on the leaf surface. More numerous than the stalked glands, the sessile glands are responsible for secretions of digestive fluid and the subsequent absorption of the products of digestion.

ABOVE: A tiny gnat (2 mm long) having landed on a butterwort's greasy leaf. Physical stimulation to the glue-tipped stalked glands causes nearby sessile glands to secrete digestive enzymes around the prey. *Pinguicula macroceras*. LEFT: Stalked glands on the leaf surface holding a droplet of mucilaginous glue. More numerous are sessile glands that are almost buried on the leaf surface. *P. primuliflora*.

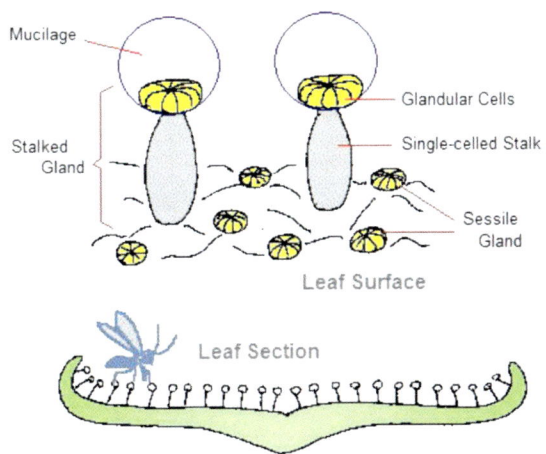

Mucilage

Glandular Cells

Stalked Gland

Single-celled Stalk

Sessile Gland

Leaf Surface

Leaf Section

Butterwort Leaf

A gnat on a leaf of *Pinguicula macroceras.*

DIGESTION

When an insect lands or crawls on a butterwort's leaf, it is mired down to the leaf surface by the mucilaginous secretions from the stalked glands. Unlike sundew tentacles, these glandular hairs offer no movement. Upon trapping a prey, nearby sessile glands begin to secrete digestive fluid. Experiments have shown that nitrogenous compounds placed on the leaf surface precipitate copious enzyme secre-

> " **Nitrogenous compounds placed on the leaf surface precipitate enzyme secretions in a matter of minutes.**

tions in a matter of minutes. The enzymes found in the secretions include amylase, esterase, phosphatase, protease and ribonuclease.

Sessile gland secretions occur only upon stimulation. It is also known that the sessile glands are capable of secreting digestive fluids only once. This is in contrast to more complex secretory systems, such as found in *Dionaea*, that can repeat the process several times. Studies suggest that this low-

viscosity digestive fluid contains a wetting agent, which allows the liquid to cover the entire insect's body more readily.

Often a trapped insect sinks down to the leaf surface — completely submerged in the digestive fluid. The secretions have been shown to contain an antiseptic substance which effectively prevents bacterial activities during the digestive process if the prey is small which, in nature, is usually the case. The resultant digested materials are absorbed through the sessile glands. Studies using radioactive isotope (carbon 14) have shown that the products of digestion are rapidly taken into the leaf tissue in a matter of hours and are carried to the other parts of the plant.

An interesting feature of *Pinguicula*'s stalked glands is that upon contact with an insect, both the gland and stalk lose turgor and quickly collapse. This helps bring the prey in close contact with the leaf surface where sessile glands abound.

Studies have shown that prey digestion in butterworts, in turn, stimulates roots to promote more efficient absorption of nutrients from soil.

Leaf Movement Often a moderate leaf movement is observed in connection with prey capture. When a small

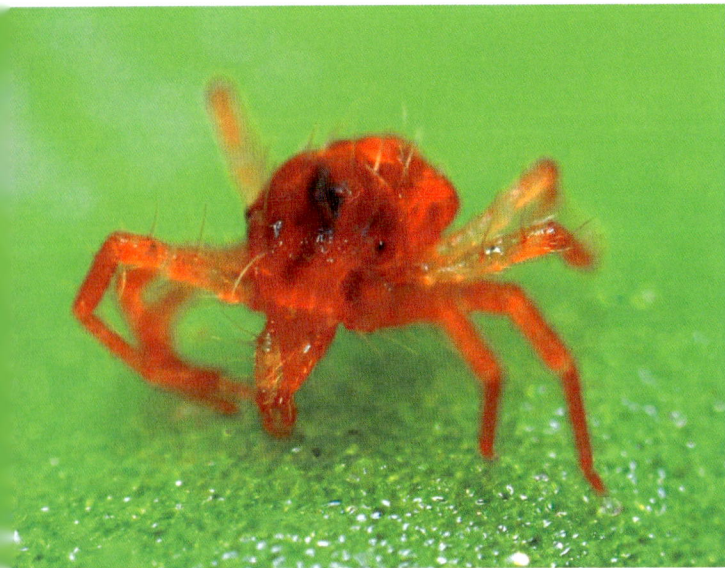

LEFT: A red velvet mite (Trombidiidae) placed on the glandular leaf surface of *Pinguicula primuliflora*. RIGHT: Within a matter of several hours, digestive juices secreted from nearby sessile glands have totally enveloped the body of the prey.

196

insect is caught on the leaf surface near the margin, the leaf begins to curl up, sometimes rather dramatically in some species. It is a slow process taking a few hours to commence and possibly lasting for several days. It is unlikely, therefore, that the leaf movement assists in prey capture as in the leaf-folding in some sundew species.

It is also noted that prey trapping away from the leaf

" Leaf curling helps keep the digestive fluids in place.

margin results in a slight dishing of the area underneath the prey. All these leaf movements are known to be caused by direct stimulation to the leaf surface and are attributed to a growth phenomenon.

Although not as fast as in sundews, the leaf movement in butterworts is a matter of common observation. It is generally believed that both leaf-curling and dishing help in holding the secretions in place during the digestion. Leaf-curling also increases the prey-leaf contact area, thus promoting more effective absorption.

TOP: A small ant (2.5 mm long) lands on a butterwort's leaf (*Pinguicula primuliflora*). The ant is immediately mired down in the tenacious glue from the stalked glands. The ant struggles to free itself but becomes more entangled. Even such athletic land animals like ants are no match for the powerful glues of the butterworts. MIDDLE: Mucilaginous glues take hold of an ant's leg. Physical stimulation to the stalked glands causes copious secretions of digestive enzymes from nearby sessile glands on the leaf surface. BOTTOM: A puddle of digestive fluids forms around the prey. Exhausted, the ant knees down as the fluids enter the victim's body. The game is over for the ant.

Pinguicula vulgaris

Pinguicula vulgaris has a wide boreal distribution both in the Old and the New World, occurring in western Europe as well as the northern part of Africa. In North America, the plant is found in the northeastern part of the U.S., the northern half of Alaska, and a large part of Canada.

The perennial, light-green rosette grows to 8 cm across, with leaf margin curling. Some plants have a discernable reddish tint on the leaf. The plants are often found on moist, moss-covered rocks, sometimes in company with sundews in the U.S. localities.

The violet flower, 1.0 cm across, blooms in June in the north-

eastern U.S. The flower scape grows to 15 cm, and then some by seed set. The upper corolla lip has two lobes. The lower corolla lip, larger than the upper, divides widely into three lobes. The lobe tips are round with no incision. Heavy, colorless hairs grow on the lower lobe surface and around the corolla tube in place of a palate. The spur is 3-6 mm long. By mid-July, the seeds set and the spherical capsule contains numerous brown seeds.

In August in the U.S. habitats, the plant forms a winter-resting hibernaculum in the rosette center made of tightly packed succulent buds.

ABOVE: A vast stretch of a *Pinguicula vulgaris* habitat on the shore of Lake Michigan, in the early morning light, in mid-June, northern Lower Peninsula. The flowers are mostly done due to somewhat early arrival of spring this year. OPPOSITE FAR LEFT: A blooming *P. vulgaris* on the shore of Lake Huron. In mid-June, the Lower Michigan Peninsula. OPPOSITE BOTTOM: A winter hibernaculum forming in the rosette center, in August. The hibernacula are made of tightly packed, succulent buds that withstands severe northern winter.

Pinguicula macroceras

ABOVE: Flowering *Pinguicula macroceras* plants along a mountain stream in southwestern Oregon, mid-April. Note a brownish leaf coloration of the plants in this colony. BELOW: *P. macroceras* on a mountain seep in southwestern Oregon, in early August.

Quite similar to *Pinguicula vulgaris*, both in leaf and flower characteristics, *P. macroceras* was previously treated as a variant of *P. vulgaris*, and its status as a separate species is still being debated. The main difference being the spur and corolla size, with *P. macroceras* being larger for both in comparison. Also, the lower corolla lobes overlap in *P. macroceras* whereas widely separated in *P. vulgaris*. The distribution of *P. macroceras* is confined to northern California, Oregon, and a few neighboring states, and Canada's Yukon Territory all the way to the southern half of Alaska. There, the distribution continues on to the Aleutian Islands, Kamtschatka and Japan.

The plants grow abundantly in some of the *Darlingtonia* sites in northern California and southwestern Oregon. These isolated populations found in the small area of California-Oregon border, exhibiting an elongated spur but no lobe overlapping, are formally described as *P. macroceras* subspecies *nortensis* — though the justification for this subspecific designation is severely challenged due to a wide overlap of characters within the range. Some plants have a strong reddish tint on the leaf in this taxon.

In northern California, the plants flower in late March into April. The violet, five-lobed flower is slightly larger, the spur slightly longer, but otherwise very much alike *P. vulgaris* flowers in the northeast. The hibernaculum-forming habit is common with *P. vulgaris*.

ABOVE: Flowers of *Pinguicula macroceras*. In mid-April, in southwestern Oregon. Colorless hairs grow around the base of the lower corolla lip in place of a palate. The violet flowers have a spur considerably longer than that of *P. vulgaris*. BELOW LEFT: The same flowers as above, showing three widely separated lobes of the lower corolla lip, without any overlap. BELOW RIGHT: *P. macroceras* flower anatomy — two stamens arching around the yellow-green, glandular ovary. The thin, frontal lobe of the stigma covers the two anthers.

The powerful glue renders an insect's desperate struggle to escape useless. Totally exhausted, the prey often collapses and lies on the leaf surface, allowing direct contact of sessile glands for efficient absorption. *Pinguicula macroceras*.

Pinguicula lutea

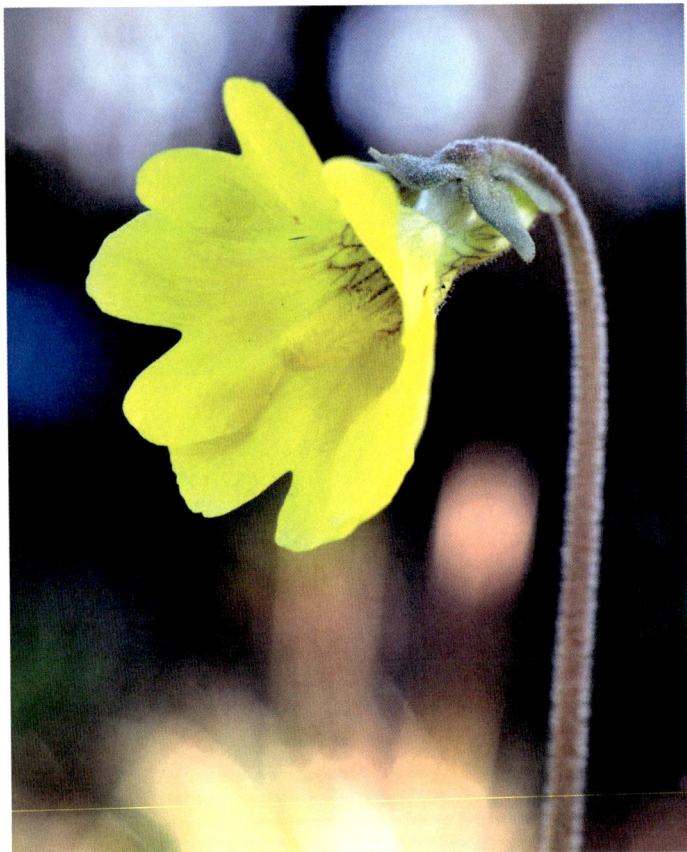

Colonies of *Pinguicula lutea* are seen along the coastal plain of southeastern U.S., from the southeastern tip of North Carolina all the way south to western Louisiana. The plant is endemic to this region and often found in large colonies, growing in moist, sandy peat on a savanna grassland.

The plant is perennial, and the rosette reaches 10 cm in diameter. The light-green leaves are strongly curled at the margins with a pointed leaf tip. The plant does not produce winter hibernacula and the surrounding grassy vegetation provides sufficient protection from the cold during winter months.

The flower blooms in early March in Florida, with a bright yellow corolla borne on a tall, glandular scape, 15-30 cm in height. The flower is large, sometimes nearing 4 cm across. Each of the five corolla lobes is notched. The lobe shape varies somewhat with occasional, additional incisions at the lobe tip, but the bilateral symmetry of the flower (including incision patterns) always seems to be maintained. The yellow, hairy palate exserts from the corolla tube, and the spur, 5-10 mm long, is yellow too. The corolla tube has brownish veins. The yellow color of the flower may be pale in some plants, and a pure white variant is also known.

The yellow flower of *Pinguicula lutea* blooming in a pendulous position, with the corolla tube showing brownish venation. In early March in Florida.

The flowers of *Pinguicula lutea* come in various corolla lobe shapes and lobe incisions. A bee is seen attending one of the flowers. The same bee moved to a nearby flower — this pollinator is a floral generalist. In early March, the Florida Panhandle.

Sharply-pointed summer leaves grow much larger than the over-wintered spring leaves. In early May, in the Florida Panhandle.

LEFT: A yellow flower bud with a glandular scape. The five-lobed calyx also carries glandular hairs. These stalked glands on the flower imply that today's adhesive-trap carnivores evolved from a "sticky" defense of floral parts against pests. In early March, Florida. RIGHT: A blooming colony in a Florida savanna, early March.

Yellow blossoms of *Pinguicula lutea* in the Florida Panhandle, early March. Swaying ceaselessly in the slightest breeze, the bright yellow flowers are among the largest in the southeastern U.S. butterworts, measuring up to 4 cm across. Serene and peaceful, this scene was captured while under fierce and relentless attack by mosquitoes.

Pinguicula caerulea

The distribution of *Pinguicula caerulea* overlaps the range of *P. lutea*, from North Carolina southwards along the Atlantic coast, but ends in the central Florida Panhandle. The plant is endemic to this region. The vegetative parts of *P. caerulea* and *P. lutea* are indistinguishable and the separation is impossible when the plants are not in flower. Both prefer constantly moist but well-drained area of the grassland.

P. caerulea is perennial with no hibernaculum formation. The rosette grows to 10 cm across. The light-green leaves are strongly curled at the margins with a pointed leaf tip.

Flowering starts in March in Florida/Georgia and continues into May. The violet flower, 2.5-3.0 cm across, has deep-purple veins. The scape grows to 25 cm and on average appears slightly shorter than that of *P. lutea*, and less stocky. The five corolla lobes have a clear notch at the tip. The lobe shapes vary from slender to round-and-overlapping. The white, hairy palate exserts from the corolla tube entrance. The spur measures 5-7 mm long. The violet corolla color varies in strength, some being very pale. An entirely white flower is also reported.

Again, *P. caerulea* and *P. lutea* are very much alike, and their distributions overlap. I have seen both flowers, purple and yellow, side by side, on a Georgia savanna in early May, but *P. lutea* flowers seem to peak a few weeks earlier than *P. caerulea* flowers.

A new flower bud is seen in the rosette center of *Pinguicula caerulea* as spring approaches. During the cold months, a rosette receives an adequate protection from surrounding vegetation in this temperate butterwort that does not produce a winter hibernaculum. In Georgia, early March.

ABOVE: Purple blossoms of *Pinguicula caerulea*, in early March, Georgia.
LEFT: The sun pierces through the violet corolla of *P. caerulea*. In late April, Georgia.

Flowering of *Pinguicula caerulea* continues into the month of May, as seeds mature in earlier flowers, with the five-lobed calyx remaining with the seed capsule. In early May, Georgia.

Reaching for the sky — a tiny butterwort, *Pinguicula pumila*, triumphantly holds a white corolla on its slender, glandular scape. In the Florida Panhandle, early March.

Pinguicula pumila

The plant is found along the Atlantic coastal plain all the way south to the Golf Coast. The distribution continues through the Florida Keys to Bahamas. The plant prefers a moist but well-drained habitat, similar to *P. lutea* and *P. caerulea*, and is often found in the same general area. *P. pumila* does not form a winter hibernaculum and may be annual in habit in North America.

This butterwort is not considered rare, but spotting one in the field may present a challenge because of its small size, sometimes even in flower. The rosette barely reaches 4 cm across at the largest end. The light-green leaves are curled at the margin.

A single plant produces many flowers in early spring. The scape grows to 10 cm tall and tiny flowers, 1.3-1.5 cm across, come in various colors even within the same colony. A white lobe color with a light yellow tube veined in reddish-brown is common, but a pale to deep-violet lobe is also seen. An entirely yellow variant is also known. A specimen with the entirely white flower is also reported from Florida. The lobes are lightly notched at the tip, and the palate not exserted. The palate color is pale yellow to white for most lobe colors but can be deep violet for violet colored lobes. The spur is 3-5 mm long.

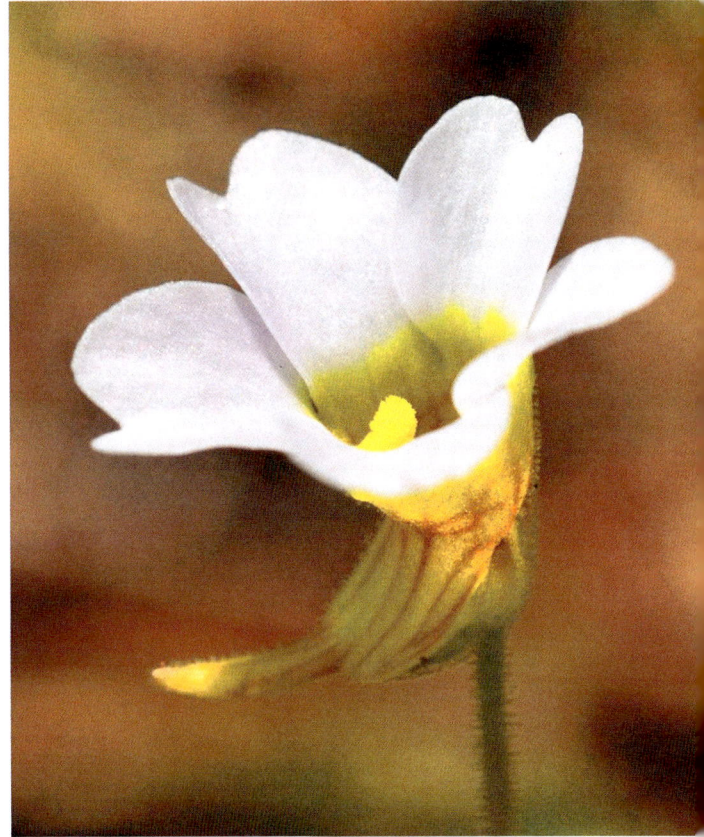

The tip of each corolla lobe has a slight notch. The yellow corolla tube has brownish veins in this flower. *Pinguicula pumila*, in May, Georgia.

BELOW: A purple-flower variant of *Pinguicula pumila,* in early May in Georgia. Often a single colony contains various corolla colors.

ABOVE & LEFT: Flower color variations of *Pinguicula pumila,* in early March in Florida.

211

Pinguicula planifolia

Limited only to the coastal plain of Mississippi and Alabama and the western part of the Florida Panhandle, this U.S. endemic is found in roadside ditches and other wet areas often with standing water. Occasional rains easily flood the area, submerging the plants temporarily. The plants growing in a drier habitat appear smaller in size in comparison.

The plant is perennial and does not form a winter hibernaculum. This is a huge butterwort with the rosette diameter reaching 20 cm across. A single leaf may grow to 12 cm on occasions. The margins of the flat leaf curl up slightly. The leaves assume a characteristic deep-maroon color (for plants growing in full sun), allowing easy identification for this distinctive species. The plants in shade may be greener though a clear reddish tint is often noticeable.

The plant blooms in early March. The tall scape, 20-30 cm long, supports a single, nodding flower. Each of the five lobes is deeply incised almost one half its length. This distinguishes this species from all the other southeastern butterworts. The corolla lobes are pale pinkish-purple, getting darker around the tube. Some flower color variation is seen and a purple corolla color is not uncommon. The lobe shape also varies somewhat from fat to slim but the deep lobe incision remains constant. The flower requires bright light to fully open, with its yellow, hairy palate strongly exserted beyond the corolla tube entrance. The spur is 3-4 mm long.

ABOVE: An unusually late blossom of *Pinguicula planifolia* seen in the central Florida Panhandle, early May. The deeply incised corolla lobe tips are sharply pointed in this specimen. BELOW ALL: The corolla color varies from almost-pure-white to bluish-purple. The lobe shape of the corolla also varies somewhat from thin to fat. The deep lobe incision, along with a strongly exserted palate, is a consistent feature for all flowers of *P. planifolia*. In early March, the Florida Panhandle.

212

The sun beams brightly on the deep-red foliage of submersed *Pinguicula planifolia* after a heavy rain. In late April, Florida.

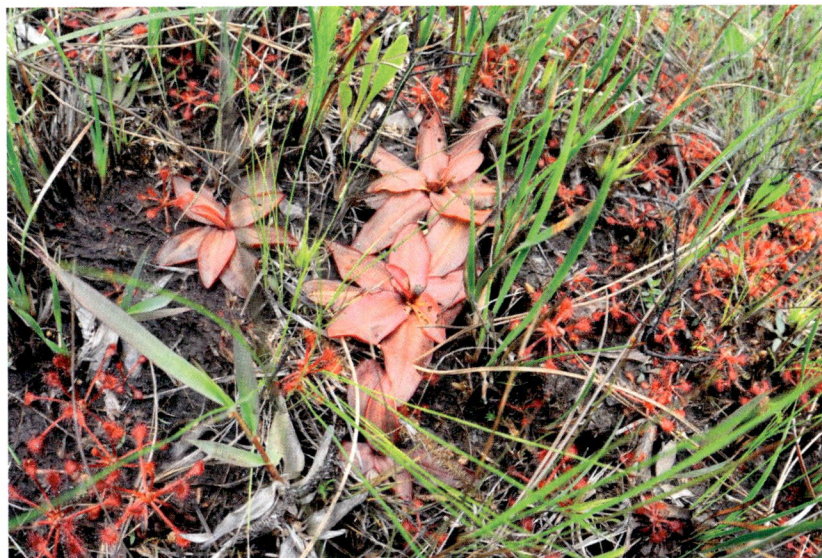

LEFT: Flowering *Pinguicula planifolia* plants growing in standing water. In early March, Florida. ABOVE: Red-leaved rosettes of *P. planifolia* sharing the moist habitat with equally red sundews (*Drosera capillaris*). One wonders who scores highest in a bug-eating contest. In early May, Florida.

Pinguicula planifolia growing in damp, peaty soil in the pit. Flowers are long gone and new leaves are sprouting, some to 10 cm or more in length. The characteristic beef-red coloration of the leaf is missing in this population. Instead, the thin foliage is assuming a pale-pinkish, translucent look. Venomous cottonmouth snakes frequent this area, I was told. In the Florida Panhandle, early May.

Pinguicula ionantha

A syrphid fly — not a true pollinator — expressing interest in the palate hairs of *Pinguicula ionantha*. The palate hairs are food bodies of the flower, cheap decoys that lure non-pollinators away from the plant's delicate reproductive parts. In early March, Florida.

Described by Godfrey *et al.* in 1961, this is the newest addition to the southeastern U.S. butterworts. The plant is only known from a small area in the central Florida Panhandle and is placed on the federally endangered species list. In spite of its extremely narrow distribution, the plant is fairly easy to spot in the region. *Pinguicula ionantha* prefers a very wet habitat subject to temporary flooding after rain, similar to a *P. planifolia* habitat. In deed, *P. ionantha* is associated with *P. planifolia* and sometimes the colonies of these butterworts are intermixed in the same general area.

The plant is a perennial forming a light-green rosette to 15 cm across. The leaf margins curl up slightly. The leaves of fully developed plants tend to be flat with only slight leaf margin curling. The plant does not form a hibernaculum.

The flower blooms in early March. The scape grows to 15 cm tall. The corolla, 2.0 cm across, has five white lobes with the purple corolla tube entrance. Each lobe has a shallow notch at the tip. The yellow, hairy palate is strongly exserted in a fully open flower in bright light. The spur is 4-5 mm long. The flower colors vary somewhat and a fairly strong purple corolla is noted in the field. In fact, both *P. ionantha* and *P. planifolia* exhibit very similar flower color variation from white to purple, as well as similar slender-to-overlapping lobe shape variation.

ABOVE: Temporary flooding easily submerses *Pinguicula ionantha* plants fond of very wet habitat. In May, the Florida Panhandle. RIGHT: The light-green rosette of a flowering specimen of *P. ionantha*. In the Florida Panhandle, early March.

Blooming *Pinguicula ionantha* in the Florida Panhandle,
early March. As in *P. planifolia*, the shape of the corolla
lobes varies somewhat from round to slender. INSERT:
A violet-flower variant. In March, the Florida Panhandle.

Pinguicula primuliflora

The plant is endemic to the Gulf Coastal region of the U.S., occurring in the western Florida Panhandle, southern Alabama, Mississippi, and Louisiana.

Pinguicula primuliflora is somewhat of a renegade among native southeastern butterworts in its habitat preference, and is unlikely to be found in the general area of coastal savannas where the others grow. The plant is fond of very wet places, and most typically found on the edges of ponds and streams, sometimes growing on a loose mat of vegetation floating on the water surface. To observe *P. primuliflora* often entails waist-deep wading in the stream — or kayaking. Curiously, this butterwort often shares the habitat with *Sarracenia rubra* ssp. *gulfensis*, which, incidentally, is also an exception among the Gulf Coast pitcher plants in its choice of habitat. *P. primuliflora* appears to prefer a slightly shady area for optimal growth.

This is a large, perennial butterwort, sometimes nearing 20 cm in rosette diameter. The leaf margins curl up moderately. The plant has an interesting growth habit of producing buds on the leaf surface, particularly near the leaf tip. This often results in a thick, clustering clump formation in the wild population. No winter hibernaculum is formed in this species.

Flowering peaks in March but sporadic blossoms may be seen throughout summer. The hairy scape grows to 15 cm to bear a nodding flower. The flower is 2.5-3.0 cm across, and is light to pale purple, turning white near the corolla tube entrance. Each corolla lobe is notched and the yellow, hairy palate exserts strongly. The spur is 4-5 mm long.

In cultivation, this is the easiest butterwort to grow, and probably most readily available among all six species indigenous to the American Southeast. Ironically, the native population is rapidly disappearing, and only several good stands are said to be remaining in the Eglin Air Force Base in western Florida.

ABOVE & OPPOSITE: Late blooming of *Pinguicula primuliflora* in the western Florida Panhandle, in early May. Colonies are formed on mossy tree trunks growing in a waist-deep pond water (infested with venomous cottonmouths). Somewhat shady condition of the habitat may have delayed the normal flowering in this site.

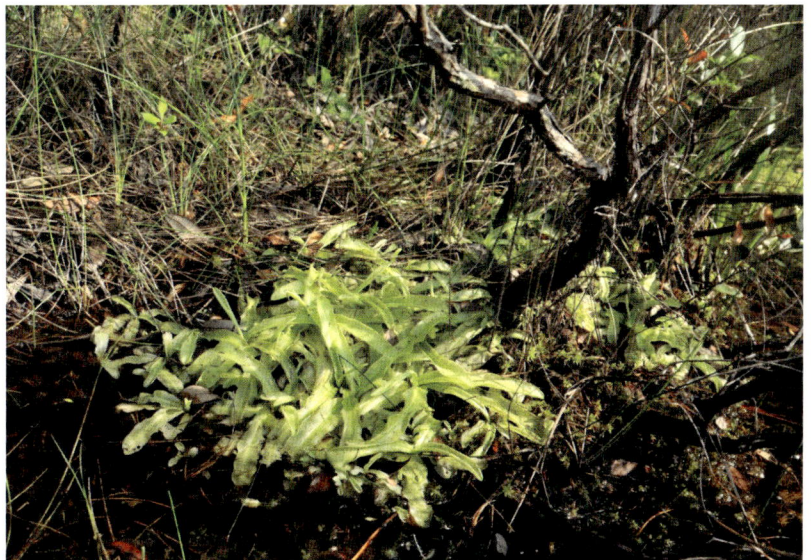

ABOVE: A clump of plants growing on a mat of floating vegetation, in early May, the western Florida Panhandle. RIGHT & FOLLOWING PAGES: Owing to its leaf-budding tendency, *Pinguicula primuliflora* often forms a dense clump of plants in a suitable habitat. This is a typical "wild" look of this species in nature, rarely reproduced in cultivation, with wavy leaves reaching 10 cm long. In early May, the western Florida Panhandle.

Bladderworts

GENUS *Utricularia*
FAMILY **Lentibulariaceae**

In the tranquil water of ponds and lakes grow plants with hundreds of tiny, balloon-like sacs attached to their branching stems. They are bladderworts, yet another kind of carnivorous plant. These sacs in the water are sophisticated miniature traps,

or bladders, designed to provide protein-rich meals for these rootless floaters. The genus name *Utricularia* is derived from the Latin word *utriculus* referring to a "small bag."

There are over 240 species of bladderworts worldwide, occurring on every continent except Antarctica, making *Utricularia* among the most diverse and successful genera of carnivorous plants. Some are terrestrial, found on moist-to-wet, often acidic soils. Others are aquatic, preferring to grow in quiet waters. A few grow in running streams attached to the rocky bottom. Terrestrial species are more commonly found in sub-tropical regions. Some species exhibit an intermediate lifestyle, capable of adapting to either terrestrial or semiaquatic habitats, depending on the level of the water table. Some tropical terrestrial species are epiphytic, found in a water-filled bromeliad rosette, with runners seeking another nearby bromeliad.

Of all the bladderworts in the world, terrestrial species are the majority, accounting for more than 80% of the genus. In evolutionary terms, aquatic species are considered more derived. Thirty-some species of bladderworts are found in the continental United States.

TOP: A bladderworts jungle. *U. inflata* (aquatic species).
OPPOSITE: Blooming *Utricularia cornuta* (terrestrial species) growing in a marl fen. In northern Michigan, mid-July.

223

"Bladderworts do not possess roots
— even in the embryonic stage.

DESCRIPTION

Bladderworts are perennial or annual herbs possessing small suction-type traps. The flower structure of *Utricularia* has a clear similarity to that of *Pinguicula*, a genus of adhesive traps in the same Lentibulariaceae family. The vegetative parts of the bladderworts, including the traps, however, show little morphological or functional resemblance to those of butterworts and construction of an evolutionary path leading to this sophisticated aquatic trap is challenging. Some butterworts sporadically produce an inwardly rolled-up adhesive leaf. One theory is that this tube-like leaf growing deep into the wet substrate and replacing the true root eventually led to the underwater traps of *Utricularia*.

The arrangement of the vegetative organs of *Utricularia* is quite peculiar in view of the general norm of flowering plants. According to P. Taylor (1989), the general morphology of the genus is described as follows: *Utricularia* does not possess roots — even in the embryonic stage — but in many (terrestrial) species structures that resemble and function as roots exist. These organs are termed *rhizoids*. A vertical stem is present only in a few species. In the majority of species, horizontal shoots called *stolons* form the framework of the vegetative part of the plant. Some aquatic species (in section *Utricularia*) sporadically produce *air shoots* arising from the stolons. These are slender, non-branching, filiform shoots which float on or toward the water surface. An air shoot is often produced in dense growth and is eventually transformed into a new plant. The true function of this structure is not known — except perhaps to aid in rapid vegetative reproduction.

Utricularia has green, flat, sometimes feather-like organs that perform photosynthesis. There are some disputes as to what to call them. "It seems reasonable to me," Taylor writes, "that plants which have, in defiance of the general rule, no radicle or roots and no true cotyledons, may be allowed to have leaves which also disobey the rules ... "

In some aquatic species, a dense cluster of leaves is formed at the growth point of the stolon at the end of the growing season. This is a *turion*, a term for winter hibernacula (resting buds) often used for aquatic species. Turions are resistant to low temperature and desiccation.

In *Utricularia,* traps arise from various vegetative organs (leaves, stolons, and rhizoids) according to the species, and positioning of the traps offers significant taxonomic characters.

In some terrestrial species (and some aquatics), the stolon develops into *tubers*. These species typically live in wet moss on rock surfaces or directly on tree trunks. The tubers, sometimes reaching 2-3 cm in length, are a water reservoir and the plants can resist temporary desiccation by withdrawing water from the tubers.

ABOVE: A sprouting bladderwort turion of *Utricularia intermedia,* in late June, southern Oregon. This species shows marked dimorphism of green leafy branches and white, trap-only branches. RIGHT: *U. inflata* floating just below the water surface, in the Florida Panhandle, late April.

Utricularia gibba in flower. Note the characteristic mat-like stolons of this species. In May, the Florida Panhandle.

INFLORESCENCE

The flowers are generally small, but quite colorful and showy, for both terrestrial and aquatic species, especially when seen in masses. Yellow is the most predominant flower color for the genus, though white to purple or bluish flowers are also common, often with yellow or reddish markings. During the flowering, which occurs from spring to late summer in the U.S. species, one often finds ponds and marshes covered with bright yellow or purple corollas. This seems to be the only time these small, obscure plants choose to announce their existence to the rest of the world.

Inflorescence is always racemose (having an unbranched flower stalk). The zygomorphic flowers have a corolla that is always two-lipped (bilabiate) having an upper lip that is generally smaller than the lower lip, but unlike temperate *Pinguicula* flowers of North America, no obvious corolla tube is present in *Utricularia* flowers. The corolla terminates in a nectar-storing spur, as in *Pinguicula*. The base of a lower lip is expanded to form a swollen palate. The edge of a lower/upper lip is either lobed or entire, depending on the species.

The arrangement of the reproductive anatomy is similar to *Pinguicula*, and is assumed to discourage self-pollination. The two stamens arch around a round ovary. The flat lobe of the stigma hangs over the anthers, with the receptive surface outside. The calyx of *Utricularia* is always two-lobed (2 sepals), except for the sections *Polypompholyx* (*U. multifida* and *U. tenella*) and *Tridentaria* (*U. westonii*) which bear a four-lobed calyx (2 normal sepals and 2 additional, small, lateral sepals).

Many species produce a reduced-size flower that self-pollinates without opening. This form of flower is called cleistogamous. In some species, cleistogamous flowers coexist with normal, open, chasmogamous flowers. In a terrestrial species, *U. subulata*, cleistogamy is quite common. In an aquatic species, *U. geminiscapa*, short-stemmed cleistogamous flowers are produced underwater along with normal flowers that protrude above the water surface.

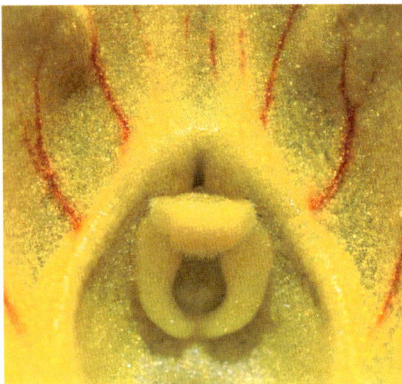

The interior of the *Utricularia* flowers with the lower corolla lip removed. Two arched stamens surround the ovary. A flat stigmatic lobe hangs over the two anthers. LEFT: *U. intermedia*. RIGHT: *U. geminiscapa*.

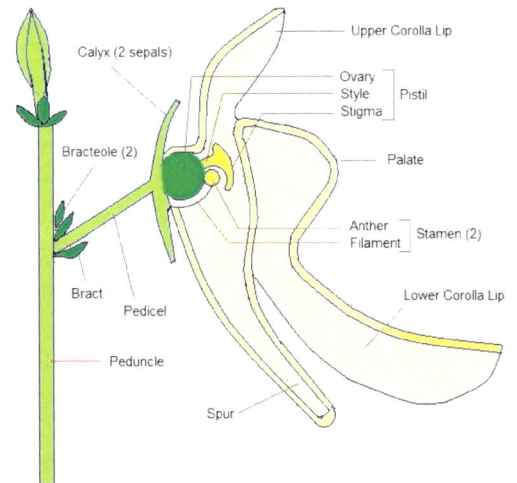

Bladderwort Flower

ABOVE: Blossoming *Utricularia intermedia* in a calm water of a sub-alpine swamp in southern Oregon, late June. Reflection of the flower on the right is showing the spur hidden under the lower corolla lip. LEFT: The yellow flower of *U. geminiscapa* with the clearly three-lobed lower corolla lip. In mid-July, southern Michigan. FAR LEFT: A hornet-mimic hoverfly seeking nectar on a *U. macrorhiza* flower. Even though the hoverfly's mouth part (proboscis) may not be well suited for the deep spur of the flower, the back of the insect's head may be touching the anthers, possibly carrying pollen to other flowers. In mid-July, southern Oregon.

SUCTION TRAP

The bladders range in size from an unimpressive 5 mm at the largest extreme to a microscopic 0.1 mm, though some Australian species occasionally produce bladders reaching 10 mm in length. The bladders are highly sophisticated mechanical traps capable of catching tiny water animals with amazing efficacy. The trap comes complete with a self-resetting mechanism. Typical prey for these miniature traps include insect larvae (especially those of mosqui-

These traps are referred to as suction traps. Prey are sucked into the bladder upon triggering the trap. The basic structure and function of the traps are common for all species — with slight differences seen between terrestrial and aquatic species. In evolutionary terms, the aquatic species (belonging to Section *Utricularia* in the genus) are considered more advanced with a complex trap door mechanism. The general description is given below for aquatic species.

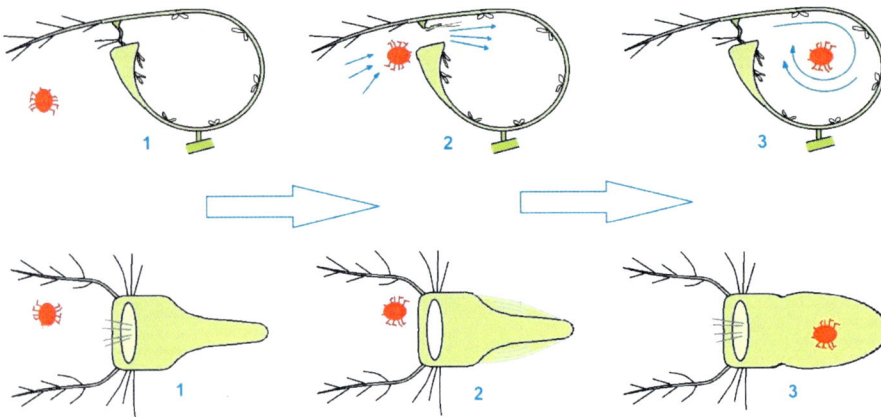

Aquatic Bladderwort Prey Trapping Sequence

Antenna There are antenna-like hairs on one side of the trap opposite the attaching stem. These hairs are non-irritable and are considered ornamental in nature. They may contribute to attracting tiny animal prey, sometimes guiding them to the trap entrance located just below the base of the hairs. The hairs may also serve to protect the entrance from flowing debris in the water.

Door The lower half of the entrance is a semi-circular valve, or a door, hinged along the upper semi-circular arc, with the free edge of the door tightly in contact with the firm collar of the lower opening of the en-

toes), aquatic worms, water ticks, and other micro-invertebrates sharing the same watery habitat. Creatures such as protozoa, rotifers and nematodes are found in the bladder of many species, sometimes as prey, but often healthy and alive in the trap.

trance, called a *threshold*. The door, which opens only inwardly, is closed and is sealed water-tight when the trap is set. When a tiny creature triggers the trap, the door flies open, swallowing the prey, as we shall see.

ABOVE: Ravenous mouths in the water, ready to devour anything that swims by. A slightest perturbation of the door is all it takes to trigger the lightning *Utricularia inflata* traps. RIGHT: A bladder turns dark purple as the prey digestion progresses inside the trap cavity. *U. inflata*.

OPPOSITE: A fully set trap on the left. Note the lateral walls are deformed in a concave shape. In the middle picture, the trap has bulged upon triggering, as the water rushes in. In mere 15 minutes or so, the trap resets itself automatically, restoring a slim, concave shape, as shown in the rightmost picture.

TRAP OPERATION

W ater is constantly being pumped out of the trap interior. Since the door is securely locked and sealed water-tight, the lateral trap walls warp inward and become concave as the water volume inside decreases. Eventually, an equilibrium is reached between the pumping force and the elastic resilience against further wall deformation. The walls cannot cave in any further, and the trap is fully set. There is a mounting water pressure from outside, but the door remains locked.

Triggering On the lower part of the outer surface of the door grow four tiny, stiff hairs (some having 6 hairs, some only 2 or none at all). These hairs function as a trigger lever. When a small aquatic creature, probably seeking a shelter in the bladderwort jungle, or perhaps lured by nectar secretions, touches one of these levers, a delicate door latch is broken. With the door now free to swing inward,

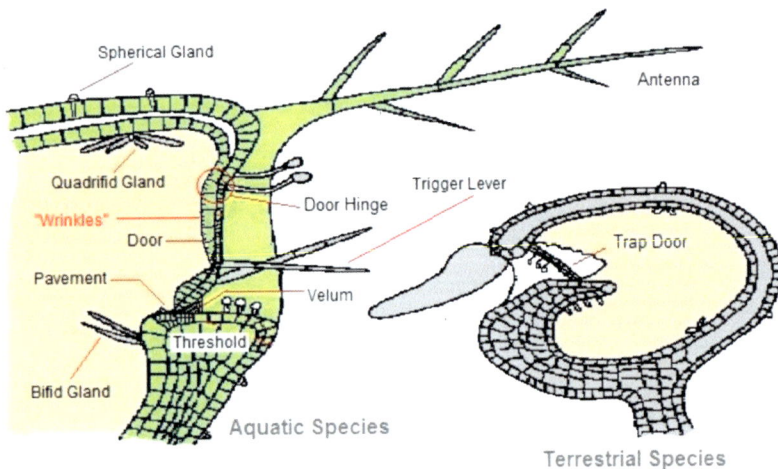

Bladderwort Trap Structure

the elastic walls bulge with the enormous force and the trap swells to its inflated shape. This causes a bladderful of water to rush into the trap, sweeping the prey with it. The door shuts back instantly, closing the entrance once again. All this happens in an astonishing 1/30-1/100 of a second. Once trapped inside, there is no hope left for the prey.

Resetting Over a period of 15-30 minutes, the trap mechanism is automatically reset as a result of the continuous pumping of water from the trap interior. The trap is now ready for another catch. In an animal-rich environment, it is not uncommon for a single trap to capture several preys until the trap cavity is completely filled.

Glands On the inner surface of the trap walls are found

two kinds of glands: two-armed glands called *bifids* and four-armed glands called *quadrifids*. The relative disposition of the projected arms of the quadrifids provides a diagnostic character of the species. The quadrifids are scattered over much of the lateral walls whereas the bifids are only found near the door (on the underside of the threshold). These different glands seem to perform specific tasks: Water is absorbed mainly by bifids while quadrifids are solely digestive in function. In addition, there are spherical glands on the trap exterior.

Water Pumping Mechanism Research has shown an active uptake of Cl- ions from the trap fluid by the glands of bifids. The osmotic gradient built by this Cl- fluxes creates movement of the water molecules, which, it is believed, ultimately drives the water expulsion from the trap. Using a trap immersed in a liquid paraffin oil, the water was observed oozing out of the trap near the door. The actual site of the extrusion is speculated to be the sessile glands covering the threshold. While many answers have been provided through research, understanding of the exact pumping mechanism and the roles the glands play awaits further elucidation.

Anoxia Research shows the trap consumes oxygen quite rapidly and the trap fluid is kept in a state of anoxia. This likely causes trapped prey to die of asphyxiation. This cellular respiration is associated with trap resetting, and it is shown that adding aerobic (oxygen) respiration inhibitors to the trap fluid severely restricts water pumping.

Digestion Over a period of several hours to a week, trapped animals become digested and absorbed by quadrifids and the nutrients are carried away to the rest of the plant. Digestive enzymes detected in the trap are believed to be secreted by the quadrifids — at least in younger traps. After the first prey is captured, bacterial actions are seen to dominate in the digestive process and the trap often assumes a dark purple color. Phosphatase is among the enzymes detected in the trap and shows the highest activity in the trap fluid's acidic environment.

Commensals Some microorganisms (such as bacteria, protozoa and rotifers) are found living in the trap fluid. These commensals facilitate the decomposition of trapped animals. In the absence of prey capture, these "free renters" seem to enjoy their stay in the trap without apparent benefits for the bladderwort.

I had just finished setting up my photomacrography gears. I looked into the viewfinder of my camera, trying to capture on the film one of nature's most intriguing trapping mechanisms in action. Against the backdrop of dark water shined two empty bladders. Then, a tiny water creature strayed into my view.

The animal swam in the water, circling around the bladders — then, as if having made up its mind, it proceeded closer and closer to the mouth of one of the bladders. As it appeared, it touched the trap door. All I could see is that the animal suddenly disappeared from the scene ...

The next thing I realized is that the animal was moving frantically inside the translucent bladder which, a moment earlier, was empty. It dawned on me that I had just witnessed a lightning-fast prey capture of a *Utricularia inflata* trap in action.

233

Marvel of Trap Setting How a tiny trap manages to exert such enormous force at prey capture is simply astonishing. This is achieved by a clever, energy-efficient scheme. The continuous respiration of the trap slowly removes water from the trap interior. This low-energy-consuming process results in a gradual buildup of mechanical stress in the trap walls. The elastic energy thus accumulated over a period of half an hour is released in a single burst upon trap triggering.

For the proper operation of the trap, the force of suction as well as the amount of water must be adequate. Water is considered incompressible, so the physical shrinkage of the trap volume, for all intents and purposes, is equal to the water volume displaced from the trap (assuming no leak).

The blasting power of the trap is a function of the wall properties. The stronger the wall elasticity, the more powerful the trap suction — provided the trap comes with a comparably strong water pump to expel the water. If the wall is flimsy, a weak pump will do the job, but the resultant suction force may not be strong enough to carry the prey into the trap. On the other hand, if you are to construct a non-collapsible trap out of titanium (like a submersible), the trap mechanism should still work, theoretically speaking. But to remove a sufficient amount of water from such a trap would require a *very* powerful pump — considering the low compressibility of water, a regional "vacuum" would form in this titanium trap!

As for the *real* traps, Nature provided a delicate balance between the bladder wall elasticity and the pumping force.

Mechanical Subtleties of the Door Lock As we have seen, the water tightness of the door is essential for the suction trap mechanism.

The surface of the threshold — against which the door edge rests — is covered with a *pavement epithelium* of sessile glands secreting mucilage. There is a slight depression in the middle of the pavement where the cells are most densely packed. The middle of the free edge of the door — which is strengthened by dense cells to make a firm edge — rests in the pavement depression. This sets a door lock.

Note that only the center of the door edge impinges on the pavement depression, with both sides of the remaining free edge of the door merely lying flat against the pavement, leaving chinks through which water can enter. To prevent the leakage, the cuticular membrane attached to the outer edge of the pavement runs along the length of the threshold, overlapping the outer edge of the free margin of the door. This thin but firm membranous tissue is called *velum*.

When the door swings back right after trap firing, it rests

"Shall we dance?"
A slender worm displays
an elegant body movement
— seemingly indifferent to
the deadly trap nearby.
Utricularia inflata.

against the velum. Any slit along the door edge creates a weak water current into the trap, which in turn pulls the thin velum toward the slit in a way to make a seal. It is believed that mucilage secreted from the stalked glands on the threshold near the door rest helps complete the seal.

The trap's internal pressure decreases as the water is continually being pumped out of the trap. This causes the door to be being pulled inward, pushing the door edge more tightly into the pavement depression. The pressure also makes the initial, outwardly-bulging door surface slightly flattened. The change in the door surface curvature of a fully-set trap is reflected in the more erect trigger lever posture. As long as the mechanical balance of the trap is undisturbed, the door remains locked water-tight in spite of the mounting water pressure outside.

Door Opening Observations of the trap operation utilizing high-speed videos have provided a new insight into the door opening mechanism. This replaces the long-held notion that the downward push of a trigger lever mechanically pulls the door edge out of the pavement depression, thus creating a small opening for water to enter. This causes the door to lose its mechanical lock altogether, forcing the door to flap open completely. To make the idea more plausible, the trigger levers grow on the lower portion of the door just around what appears to be a secondary hinge. This makes it possible to unlatch the tight lock with a minimum of stress, the old theory speculates.

The new idea embraces "buckling" as the key mechanism for door opening. Buckling is a well-studied physical phenomenon and is thoroughly analyzed mathematically. Buckling is characterized by a sudden change of structural states under an increasing load — in bladderworts, a flip of the trap door curvature upon triggering, from convex to concave, as seen from the outside of the trap.

When the trap is set, the door is bulging outward. This surface curvature allows the door to withstand a strong outside pressure while a delicate structural equilibrium is maintained. If a trigger lever is touched, the surface area at the base of the lever is disturbed. This slightest perturbation of the door surface — under a near-critical pressure — causes the door to buckle. The buckling starts where the trigger lever grows and propagates swiftly to a larger area, reversing the curvature of the entire door surface. As the buckling reaches the free edge of the door, the change of the angle of the door edge toward the pavement depression virtually unlocks the door.

Giving in to the enormous pressure from outside, the door opens in a matter of 1/1000 second. The sudden inrush of water forces the door to be held in the open position. Water continues to flow into the trap, carrying the prey with it. The elastic energy of the trap walls now released and the trap fully inflated with no structural stress, the door swiftly snaps back to the closed position, unbuckled.

It is clear that any disturbance to the trigger lever will cause buckling — however, the trigger lever may not be necessary for trap triggering. In fact, some traps do not have trigger levers, and some has only stalked mucilage glands in place of trigger levers (like *U. purpurea*). In reality, a more likely scenario would be that a prey animal bumps onto the door itself, which will surely create enough surface perturbation to cause buckling. For the majority of aquatic traps that do possess trigger levers, they might just be largely ornamental.

Perhaps the most critical structure for trap door buckling resides in the middle portion of the door where the wall becomes thinnest and most touch-sensitive — and appears to function as a secondary hinge. This is a vulnerable spot of the door and a slightest push by a potential prey will cause buckling (like your knee buckles if pushed from behind). Not coincidentally, this is where the trigger levers are located.

The main purpose of the levers might be to allure and direct a potential prey to this "sweet" spot of the door. The double-layered construction of the trap door with the much thicker inner layer — with wrinkles to allow for expansion — and this secondary hinge mechanism appear to hold the secret to "an astounding degree of mechanical delicacy … "

Besides pumping of water by the wall glands (which is physiological in nature), setting, tripping, and resetting of the trap are believed to be purely mechanical, unrelated to growth phenomena. Therefore, one bladder can repeat the triggering action without any biological growth limitation. One observer counted 24 times of trap resetting.

Spontaneous Firing Research shows the water pumping continues (after trap resetting) for as long as several hours. A recent study has also revealed that, in the absence of prey-induced triggering, a trap may fire spontaneously — often at some constant rate — without any apparent external stimuli. This seems to be a general phenomenon for any trap.

Spontaneous firings seem to occur as the pumping force causes the pressure to reach a critical buckling threshold. It is unclear what benefit, if any, the spontaneous firing may bring to the plant, and whether it is merely a (mechanical) consequence of over-pumping or it has some metabolic significance.

The frequency of spontaneous buckling may depend on the door rigidity — the turgor pressure of the door influencing the pressure tolerance for buckling.

Is the door triggering really just purely mechanical? If a physical stimulus to a trigger lever somehow causes the turgor pressure to weaken, immediate bucking is unavoidable. If true, this has a physiological implication, though research to date has not detected any action potentials during trap triggering.

Utricularia gibba

The plant has a worldwide distribution. In North America, *Utricularia gibba* is found in the northern U.S. and Canada. The plant carries a complex taxonomic history because of its wide variation in flower as well as vegetative parts. *U. biflora* is a now-invalid epithet referring to the 2-flowered variant of *U. gibba*.

U. gibba is an affixed, aquatic perennial, growing in waters of various depths but flowering seems to occur only in plants in shallow waters a few centimeters or less. The green stolons often get interwoven like a mat. The traps are 1.0-2.5 mm long. The plant forms a winter turion.

Massive flowering is often seen in roadside ditches, sending up slender peduncles, 10-20 cm high, each bearing a few flowers. The flower is bright yellow, up to 2 cm long, often with red streaks on the swollen base of the lower corolla lip. The upper corolla lip may be weakly three-lobed and as large as or slightly larger than the lower lip. The spur is as long as the lower lip.

ABOVE: Profuse blossoming of *Utricularia gibba* in Georgia, early May. BELOW: A close-up view of the flowers. The upper corolla lip in *U. gibba* grows as large as the lower lip.

Utricularia striata

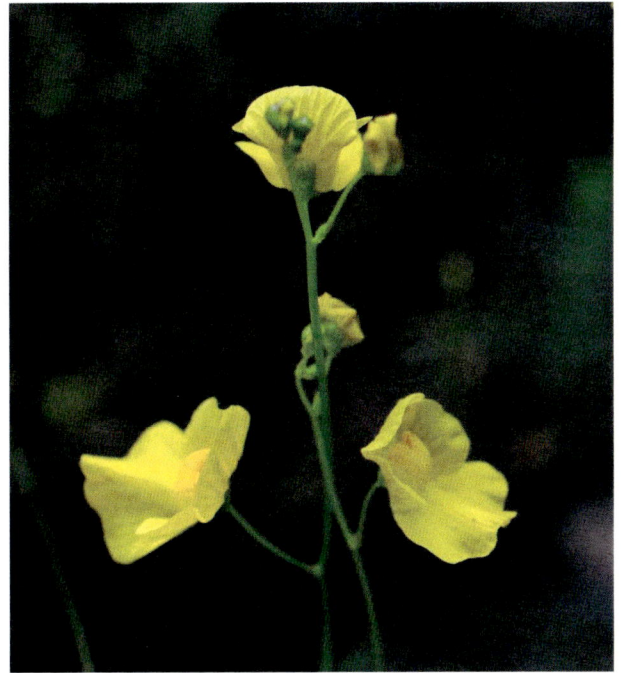

Endemic to the eastern coastal plain of the U.S., the plant is an affixed, semiaquatic perennial, occurring in shallow water and on wet soil. The plant is dimorphic, and the small, leafy stolons, 4 cm across, 5-10 cm long, with sporadic traps grow in water, while the white, trap-only stolons grow into the substrate that bear numerous traps, 1-2 mm long. The plant forms a winter turion. *Utricularia striata* was formerly known as *U. fibrosa*.

The plant produces yellow flowers, 1.5 -2.0 cm long, on a tall peduncle 10-30 cm long, with red streaks on the swollen base of the lower corolla lip. The upper corolla lip is shallowly three-lobed and as large as or slightly larger than the lower lip. The spur is as long as the lower lip. The general appearance of the flower is very similar between *U. gibba* and *U. striata* and both species flower well on wet soil or in very shallow water. The distinction requires investigation of the vegetative parts.

ABOVE: Flowers of *Utricularia striata* resemble those of *U. gibba*. In the western Florida Panhandle, early May. BELOW: *U. striata* growing in wet, peaty soil on the lake margin, early May, Florida.

Utricularia floridana

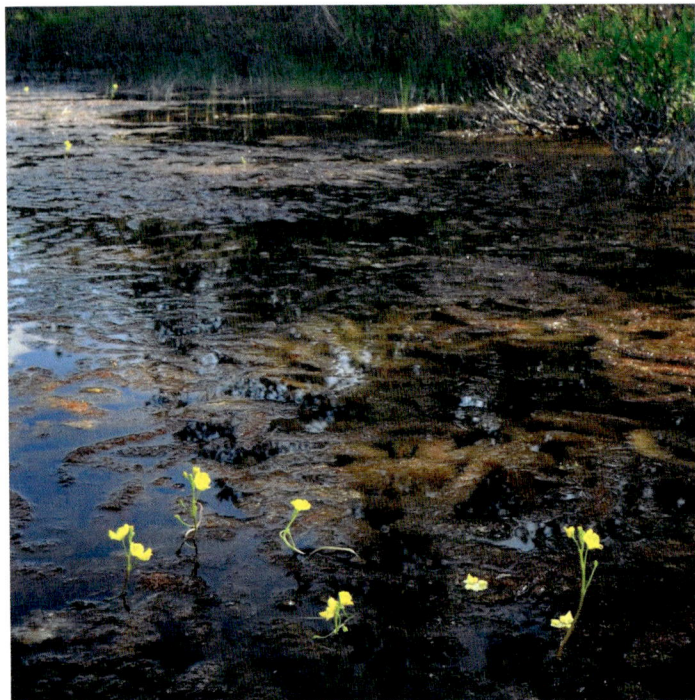

Found in Florida and other neighboring states, this is a large and imposing U.S. endemic. *Utricularia floridana* is an affixed, aquatic perennial, typically found in deep waters of ponds and lakes. The plant exhibits clear dimorphism and produces white stolons only bearing numerous traps, in addition to the main, green stolons having leafy branches with only sporadic traps. The trap-only stolons grow in the substrate and totally lack in chlorophyll, with traps being 1.5-3.0 mm long. The green, leafy stolons, 50 cm long and 8 cm across, grow in water, often meandering toward and near the water surface. These leafy stolons are reminiscent of a fluffy "foxtail." The plant forms a winter turion.

In May, the plant produces yellow flowers 2.0 cm long, often with reddish streaks at the swollen base of the lower corolla lip. The upper corolla lip is shallowly three-lobed and characteristically larger than the lower lip. The cylindrical spur is slightly shorter than the lower lip and often hidden under the lower lip.

Firmly affixed at the lake bottom, this is the only bladderwort species capable of penetrating a meter-deep water and bearing flowers in the air. The robust peduncle (to 3 mm thick) sometimes reaches 1 m or more, protruding the upper, flower-bearing portion (10-15 cm from the tip) above the water surface. Each peduncle bears 5-20 flowers.

ABOVE: *Utricularia floridana* in flower, in a pond in the western Florida Panhandle, early May. BELOW: Leafy, photosynthetic stolons in the water bear only sporadic traps. *U. floridana* also produces white, chlorophyll-free, trap-only stolons that descend deep into the substrate at the lake bottom. In early May, Florida.

In Florida, *Utricularia floridana* flowering starts in late April and continues into May, though yearly fluctuation is common. The large flowers are borne on a peduncle, 2-3 mm thick, and one meter long. The upper corolla lip is clearly larger than the lower lip and is shallowly three-lobed. In early May, the western Florida Panhandle.

Utricularia floridana is a large and imposing species. In an ideal habitat, a colony of the plants forms a breath-taking underwater forest of "foxtails," covering the pond with green, leafy stolons meandering near the water surface. In early May, the western Florida Panhandle.

Utricularia macrorhiza

The plant occurs in North America and eastern temperate Asia. In North America, the plant is found in the northern part of the U.S. and Canada. *Utricularia macrorhiza* is a suspended aquatic perennial, with traps 1.5-5.0 mm long. It has been treated as *U. vulgaris* until recently when P. Taylor (1989) described the North American plants as a separate species. The plant forms a winter turion.

The plant produces yellow flowers in June (in the U.S. habitats), 1.5 -2.0 cm long, with reddish-brown streaks on the base of the swollen lower lip. The lower corolla lip is larger than the upper, with its margin being entire. The cylindrical and curved spur is as long as the lower lip. The tall peduncle grows to 40 cm long, but only a portion of the peduncle protrudes above the water. Each peduncle bears several flowers. The green, leafy stolons typically float near the water surface, but seem to sink deeper for flowering plants.

ABOVE: The long peduncle of *Utricularia macrorhiza* protruding above the water, in a swamp in southern Oregon, mid-July. BELOW LEFT: *U. macrorhiza* assuming a reddish-purple coloration, in a pond in northern Michigan, in July. BELOW RIHGT: A floating plant of *U. macrorhiza* in southern Oregon, mid-July.

The inflorescence of *Utricularia macrorhiza*, in a swamp in southern Oregon, mid-July. Note a pointed, curved spur under the lower corolla lip.

Utricularia macrorhiza.

Utricularia intermedia

The plant has a circumboreal distribution throughout North America, Europe and Asia. The plant is an affixed aquatic perennial, found on wet soil and in water to 40 cm deep. *Utricularia intermedia* is often found growing in company with *U. minor*. The plant exhibits marked dimorphism. The chlorophyll-free stolons, to 20 cm long, bearing numerous traps, anchor firmly in the substrate. The white traps are often as large as 5 mm in length. The green, leafy stolon, up to 30 cm long, that bears no traps, ascends in the water toward the surface, forming what can be described an "underwater forest." In a very shallow water, the bright green leaves of the plant cover the soil surface. The plant forms a winter turion toward the end of summer. The round turion, 1 cm across, sprouts in mid-June next season.

In northwestern U.S. habitats, the plant flowers in July. The deep-yellow flowers, 1.5 cm long, are borne on a peduncle 20 cm long, a few flowers per peduncle. The lower corolla lip, much larger than the upper lip, has a rounded swelling at the base with red streaks, and is entire. The spur is slightly shorter than the lower lip.

A spectacular colony of *Utricularia intermedia* formed in a shallow water (20-30 cm deep) of a sub-alpine swamp, at an elevation of 5800 feet. In mid-July, southern Oregon.

LEFT: A sprouting winter turion of *Utricularia intermedia*. Note the marked dimorphism of green, leafy stolons and white, trap-bearing stolons that penetrate into the substrate. In late June, southern Oregon. BELOW: Seed capsules of *U. intermedia*. Inserted at the base of the pedicel is a bract with accompanying bracteoles.

ABOVE: The trap of *Utricularia intermedia*. Note that the trap contains no green chlorophyll pigments because in this dimorphic, aquatic species, the trap grows in the soil, as in terrestrial species. For this very reason, the trap structure resembles that of terrestrial bladderworts.

LEFT: The bright-yellow flowers of *Utricularia intermedia* blooming in a sub-alpine swamp. The edge of the lower corolla lip is entire (not lobed). In southern Oregon, mid-July.

245

ABOVE: *Utricularia intermedia* growth habit with a thin surface-water covering. In this species, the traps are always buried deep in the substrate and never exposed on the soil surface or in the water. In mid-July, southern Oregon. LEFT: Damselflies are quite common in this sub-alpine swamp. A moment after I took the picture, this one flew into the air and collided with a moth, or so it appeared. Wrong! It captured the moth in the air, landed on grass nearby, and started to devour the struggling moth, head first. In mid-July, southern Oregon.

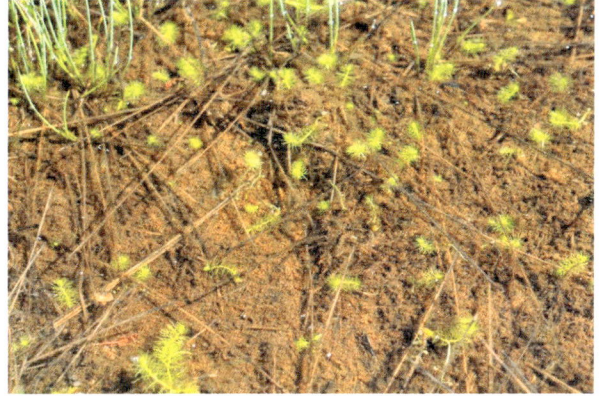

RIGHT THREE: Sprouting winter turions of *Utricularia intermedia* as water temperature rises. The turions remain in the substrate as the new green leaves emerge. In the third picture an emerging turion was pulled from the substrate to show white traps. In late June, southern Oregon.

BELOW: *Utricularia intermedia* growth habit in a relatively deep water (up to 30 cm deep). Foxtail stolons meander toward the water surface. In late July, southern Oregon.

Utricularia minor

The plant has a circumboreal distribution. In North America, *Utricularia minor* is often seen with *U. intermedia*. The plant is usually an affixed aquatic perennial, found in very shallow to deep waters. The stolon is dimorphic, albeit not consistently so. Sometimes the green stolon bears both leaves and traps and trap-only stolons may not be present. The stolon sometimes assumes a strong reddish coloration. The tip of the stolon

ABOVE: *Utricularia minor* in flower, in southern Michigan, mid-July.

RIGHT: The pale-yellow flower of *Utricularia minor,* in mid-July, southern Michigan.

Sprouting winter turions of *Utricularia minor,* in late June, southern Oregon. Note the wide, reddish leaves.

The traps are 0.8-2.5 mm long. The plant forms a winter turion. The turion sprouts in June in the U.S. habitats.

The plant bears small, pale-yellow flowers, 1.0 cm long, on a peduncle 15 cm tall. The small upper corolla lip is ovate. The lower lip is longer than wide, and the downward side-folding of the margins makes the lower lip appear even narrower. The short spur hides beneath the lower lip.

BELOW: *Utricularia minor* growing in shallow water of a marl fen, northern Michigan, in mid-July. Stolon dimorphism is not clear in this growth.

Utricularia geminiscapa

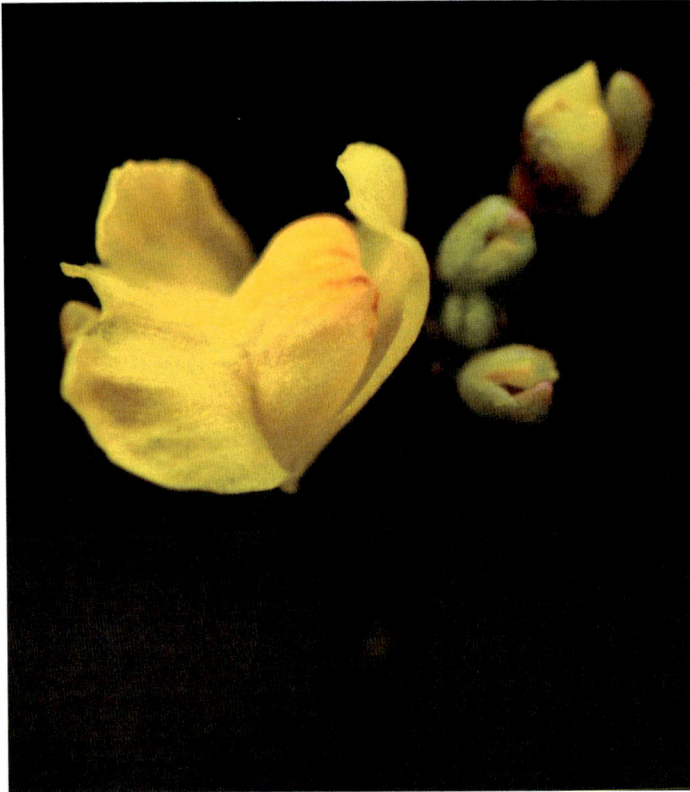

Endemic to North America, this free-floating, aquatic perennial occurs in the northeastern U.S. and eastern Canada, but is somewhat infrequent in sighting. *Utricularia geminiscapa* may be dimorphic, and in addition to the main green stolons having leafy branches (with some traps), trap-bearing stolons are sometimes noticeable. The traps are 0.6-3.0 mm long. The plant forms a winter turion.

Many peduncles, 5-15 cm long, each supporting a few flowers, are produced along the main stolon reaching half a meter in length. The corolla is yellow, measuring 6-8 mm long. The upper corolla lip is smaller than the lower lip. The lower lip is clearly three-lobed and has a rounded swelling at the base, with some reddish streaks. The cylindrical spur is shorter than the lower lip.

A distinctive characteristic of this species is the production of cleistogamous flowers. From the same branching node where a regular, aerial peduncle grows, one or more short branches emerge in the water bearing a round, closed bud at the tip. These flower buds (remaining underwater) never open but self-pollinate to bear viable seeds.

Sometimes long, non-branching segments are produced from the main stolon in dense growth. These structures are termed air shoots, and they eventually develop into individual plants.

The yellow aerial flower of *Utricularia geminiscapa*. The three-lobed lower corolla lip is larger than the upper lip. In mid-July, southern Michigan.

LEFT & MIDDLE: The cleistogamous flower in the water formed at the tip of a short branch. RIGHT: A trap of *U. geminiscapa,* in mid-July, southern Michigan.

ABOVE: *Utricularia geminiscapa* with numerous traps, in southern Michigan, mid-July. BELOW: Branching stolons of *U. geminiscapa* floating on the water surface, in southern Michigan, early July. RIGHT: Flowering *U. geminiscapa* in habitat. Southern Michigan, in mid-July.

Utricularia inflata

Endemic to the southeastern coastal plain of the U.S., the plant is a suspended aquatic perennial. The green, leafy stolons bear traps 1.0-3.0 mm long. *Utricularia inflata* does not form a winter turion, but can withstand the cold being frozen solid during the winter months. The plant is known to produce air shoots in dense growth. An air shoot is a long, slender, leafless stem at the tip of which forms a small bulge. This is actually a bud which, in a few weeks, grows into a new plant. Perhaps an airshoot is a *Utricularia*'s vegetative reproduction mechanism. It is also observed that in a stressed condition where the water level becomes extremely low, the stolon of the plant becomes swollen to form a tuber (in order to retain water).

An interesting feature of the inflorescence of this species is that a wheel of radiating, spoke-like floats is formed at the mid-point of its peduncle 20-25 cm long. This functions as a flotation device to provide a stable support for flowers above the water. The wheel, 10-15 cm across, usually has 6-8 floats. There are only a few bladderworts in the world (two in North America) that develop this type of flower flotation mechanism.

Each peduncle supports 4-17 deep-yellow flowers, 2.0-2.5 cm long. The lower corolla lip is deeply three-lobed and larger than the upper lip, which is sometimes shallowly two-lobed. The base of the lower lip is swollen. The narrowly cylindrical spur is shorter than the lower lip, and its apex is bifid.

" A flotation device affords a stable support for tall peduncles.

The wheels of floats of *Utricularia inflata* covering the pond surface. In late April, the Okefenokee Swamp, Georgia.

LEFT: The deep-yellow flower of *Utricularia inflata*, with a clearly three-lobed lower corolla lip. In Georgia, early May. BELOW: Supported by a wheel of floats, yellow corollas of *U. inflata* inflorescences reflect on the dark, tannin-colored water of the Okefenokee Swamp. In Georgia, late April.

Air shoot development sequence, from left to right.

Utricularia inflata sporadically produces a small, round organ at the tip of a long, non-branching segment which, in time, develops into a new plant. This slender segment is termed an air shoot. Air shoots are often produced in dense growth.

253

Utricularia radiata

Closely related to *Utricularia inflata*, but having a broader distribution, the range of this North American endemic extends westward to eastern Texas and northward to Nova Scotia. *U. radiata* is a suspended aquatic and appears annual with no winter turion formation. The green, leafy stolons bear traps 0.7-2.0 mm long.

The plant produces yellow flowers, 0.9-2.0 cm long. Each peduncle bears 3-4 flowers. The flower has its lower corolla lip deeply three-lobed. The upper lip is smaller. The cylindrical spur is considerably shorter than the lower lip, and its apex is not notched.

The flowers are supported on a flotation device similar to that of *U. inflata*, but smaller, with the float wheel 5-8 cm across. The float is formed at the point several cm from the tip of the peduncle 10-15 cm long. Typically, there are 5-7 floats per wheel. The overall shape of the float differs slightly between *U. radiata* and *U. inflata*.

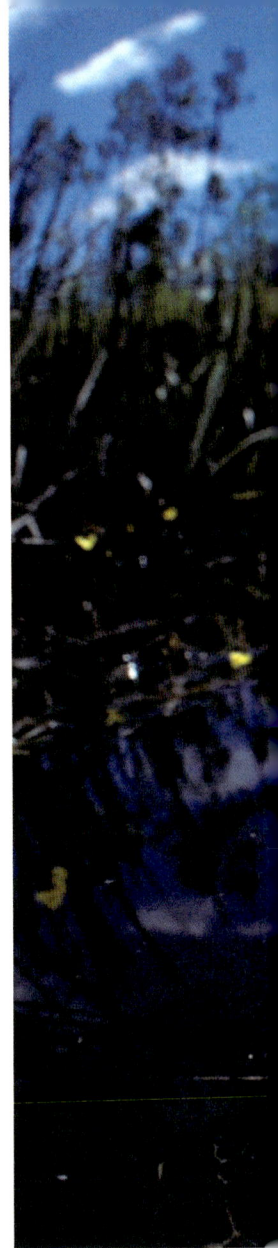

A flotilla of tiny floats of *Utricularia radiata* crowds the surface of a small pond in the Florida Panhandle, in early May. The wind had pushed the floats to one side of the pond. This is where I had a close encounter with a huge cottonmouth some years ago. Also known as the water moccasin, this highly venomous, semiaquatic pit viper suddenly appeared just a few feet away from me in the pond, stretched its upper body above the water, and yawned with its white mouth wide open. The blood in my entire body froze.

" Miniature floaters are reminiscent of children's toys left in a pool.

ABOVE: *Utricularia radiata* flowers floating on a calm water reflecting the blue Floridian sky. *U. radiata* closely resembles *U. inflata* — only a slightly reduced version. The delicate flowers supported on a float remind one of children's toys floating in a pool. In late April, the Florida Panhandle.

LEFT: The lower corolla lip of *Utricularia radiata* is clearly three-lobed, as in *U. inflata*. The floating wheel is smaller, typically around 5 cm across, and has 5-7 radiating floats.

255

Utricularia purpurea

Occurring in North and Central America and Antilles, this free-floating, aquatic perennial (probably) prefers deep waters. In the U.S., the plant is found in the Atlantic coastal plain all the way down to the Gulf Coast. The reddish vegetative parts, along with stolons having characteristic whorls of branching segments, identifies this species fairly easily. The growth tip of the stolon has curly leaves. The plant is not known to form a winter turion. The traps are 1-2 mm long. In *Utricularia purpurea*, the trigger levers are replaced with gland-tipped hairs. The carnivorous nature of this bladderwort is being questioned.

The purple flower makes the identification complete, though corolla color varies somewhat, from deep pink to pale purple. (An entirely white flower is reported in New Hampshire.) Some flowers have a yellow blotch on the base of the lower corolla lip. The flowers are 1-2 cm long, 2-3 per peduncle. The lower corolla lip, larger than the upper lip, is deeply three-lobed with lateral lobes being tube-like. The conical spur is shorter than the lower lip.

Blossoming *Utricularia purpurea* plants, growing in an alligator-infested water of the Okefenokee Swamp. In early May, Georgia.

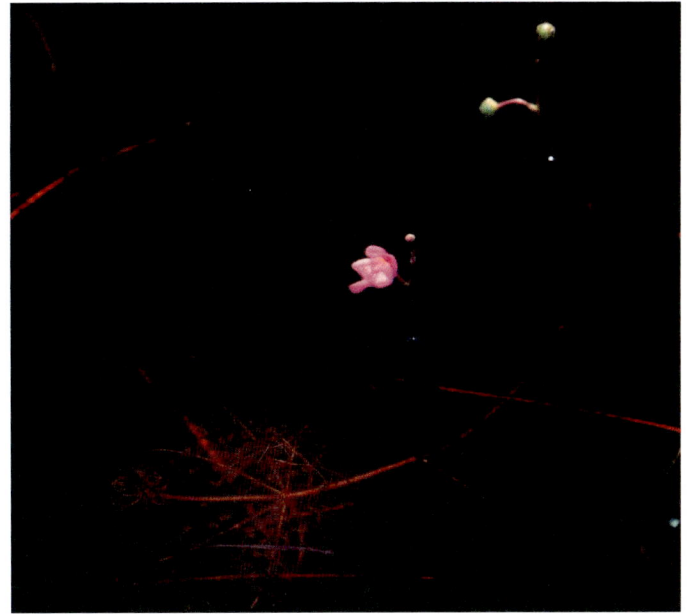

ABOVE LEFT: A deep-purple flower of *Utricularia purpurea*, in July, North Carolina. RIGHT: Removing the surface reflection reveals the whorls of branches underneath the water surface, bearing numerous bladders.

BELOW TWO: *Utricularia purpurea* blooming in a roadside ditch. In late July, North Carolina.

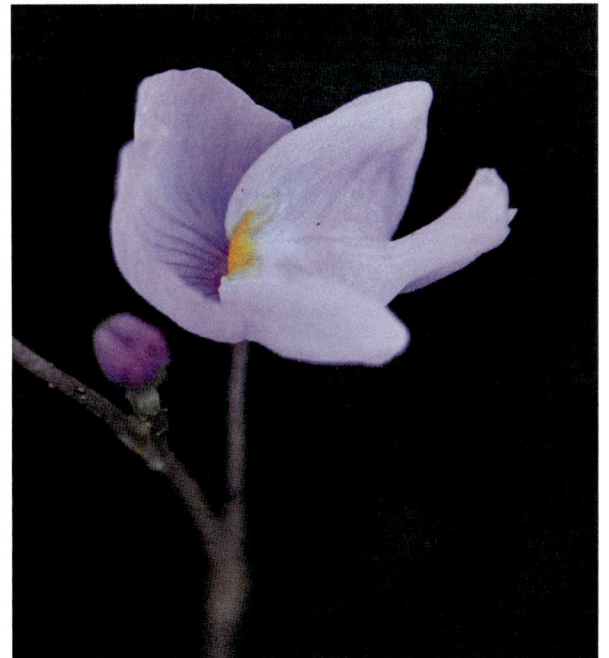

A pale-purple flower of *Utricularia purpurea* with a yellow blotch on the lip base. In the western Florida Panhandle, early May.

Utricularia subulata

This species has a pan-tropical distribution and is the most widespread among all *Utricularia* species in the world. The plant is an annual terrestrial and forms no winter turion. The stolons are white and several cm long. The green, narrow leaves are up to 2 cm long. The minute traps, 0.2-0.7 mm long, appear on stolons and leaves.

Being ubiquitous in many carnivorous plant habitats in the U.S., the plant produces numerous bright yellow flowers, 1.0 cm long, on a characteristically zigzag peduncle, 10-20 cm tall. The flower color may be pale yellow in some. The peduncle bears bracts at the branching points. The lower corolla lip is larger than the upper lip, and is three-lobed. The narrow spur is as long as or sometimes longer than the lower lip. The plant produces cleistogamous flowers that are round, closed buds capable of producing viable seeds by self-pollination. The cleistogamous flowers appear on the same peduncle with normal, open, chasmogamous flowers.

ABOVE: The bright-yellow *Utricularia subulata* flower. Early May, in Florida. RIGHT: A whole plant of *U. subulata*. Note that the traps occur on stolons as well as on leaves.

ABOVE: *Utricularia subulata* growing on a moist, white sand surface. In late July, North Carolina. BELOW LEFT: The flower of *U. subulata* on a zig-zag peduncle. In May, in Florida. BELOW RIGHT: *U. subulata* growing in peaty soil, in Florida, early May.

259

Utricularia cornuta

The plant occurs in North America, the Bahamas and Cuba. *Utricularia cornuta* is closely related and very similar in appearance to *U. juncea*. In North America, *U. cornuta* occurs much farther north and farther west than *U. juncea*. The plant is probably a perennial terrestrial, or semiaquatic, with white stolons several cm long and green leaves 1 cm long. The minute traps, 0.3-0.8 mm long, grow on stolons and leaves. *U. cornuta* forms no winter turion. The plants are often seen blooming in masses in peat bogs in the south and in marl fens in the north.

The flower is deep yellow, 1.0-1.5 cm long, borne on a peduncle 10-20 cm long. The flowers of this species tend to cluster near the tip of its peduncle. The flower has a lower corolla lip that is much larger than the upper lip. The lower lip is broadly rounded, with the free margin being entire. The tapering spur is slightly curved and points downward.

LEFT: A bumble bee moving from one *Utricularia cornuta* flower to another, in a marl fen in northern Michigan, mid-July. With the weight of the bee pushing the corolla open, the long tongue (proboscis) of a bumble bee can easily lap up nectar accumulated in the flower's spur without pushing its head deep into the corolla interior. The pollen is likely to be deposited on the bee's proboscis. BELOW LEFT: A seemingly endless expanse of bright yellow blossoms of *U. cornuta* in a marl fen on the shore of Lake Huron. In mid-July, in the northern tip of the Lower Peninsula of Michigan. BELOW RIGHT: The flower interior of *U. cornuta* with the lower corolla lip removed. Two stamens stand holding the anther. A thin stigma lobe overhangs the anthers.

Utricularia cornuta flowering in a marl fen on the shore of Lake Huron. In mid-July, northern Michigan. The plants grow and bloom well in a shallow surface water. The flowers tend to cluster at the top of the peduncle in this species.

ABOVE: A leaf with a trap (top) and a stolon bearing traps (bottom). *Utricularia cornuta* in northern Michigan, July. RIGHT: *U. cornuta* leaves in water-logged peaty soil. In southern Alabama, early May.

261

Utricularia juncea

The plant occurs in North and Central America and northern South America. In the United States, this species grows in the eastern coastal plain from New Jersey to Texas. The plant is a terrestrial perennial, with white stolons several cm long and green leaves 1.5-2.0 cm long. The traps measuring 0.3-0.8 mm long grow on stolons and leaves. The plant forms no winter turion.

The flower morphology is quite similar to that of *Utricularia cornuta*. The yellow flower, 0.5-1.0 cm long, is slightly smaller than *U. cornuta* flowers. The lower corolla lip is larger than the upper lip, and is broadly rounded, with the free margin being entire.

The peduncle grows to 30 cm, appreciably taller than that of *U. cornuta*. The flowers are more spaced near the top compared with tight clustering at the tip of the peduncle in *U. cornuta*. In *U. juncea*, seasonal cleistogamous flowers are observed in the same peduncle. The long spur tapers and points downward.

Utricularia juncea, growing in sandy soil, in late July, North Carolina. The Yellow flowers tend to be more spaced at the top of the peduncle, compared with the heavy clustering of *U. cornuta* flowers at the peduncle tip. The hair-like inflorescences seen on the ground are those of *U. subulata*.

262

In closing this account one cannot but wonder at the astonishing variety of trap structure. It is not less astonishing that there is no evidence that one form of trap is superior to another in action. The fact of variety is one with the same phenomenon observed when we survey attentively some other unit of structure. It seems as though nature, or to deify her fruitfulness, Nature, is not nor ever has been content to make some one thing, however satisfactory, and to let it go at that. She must show that she is not bound to the details of a pattern that, in this case, she can make a whole shelf full of different kinds of traps, as if to puzzle you to pick the best.

— F. E. Lloyd

The Carnivorous Plants (1942). Chapter XIV, The Utricularia Trap.

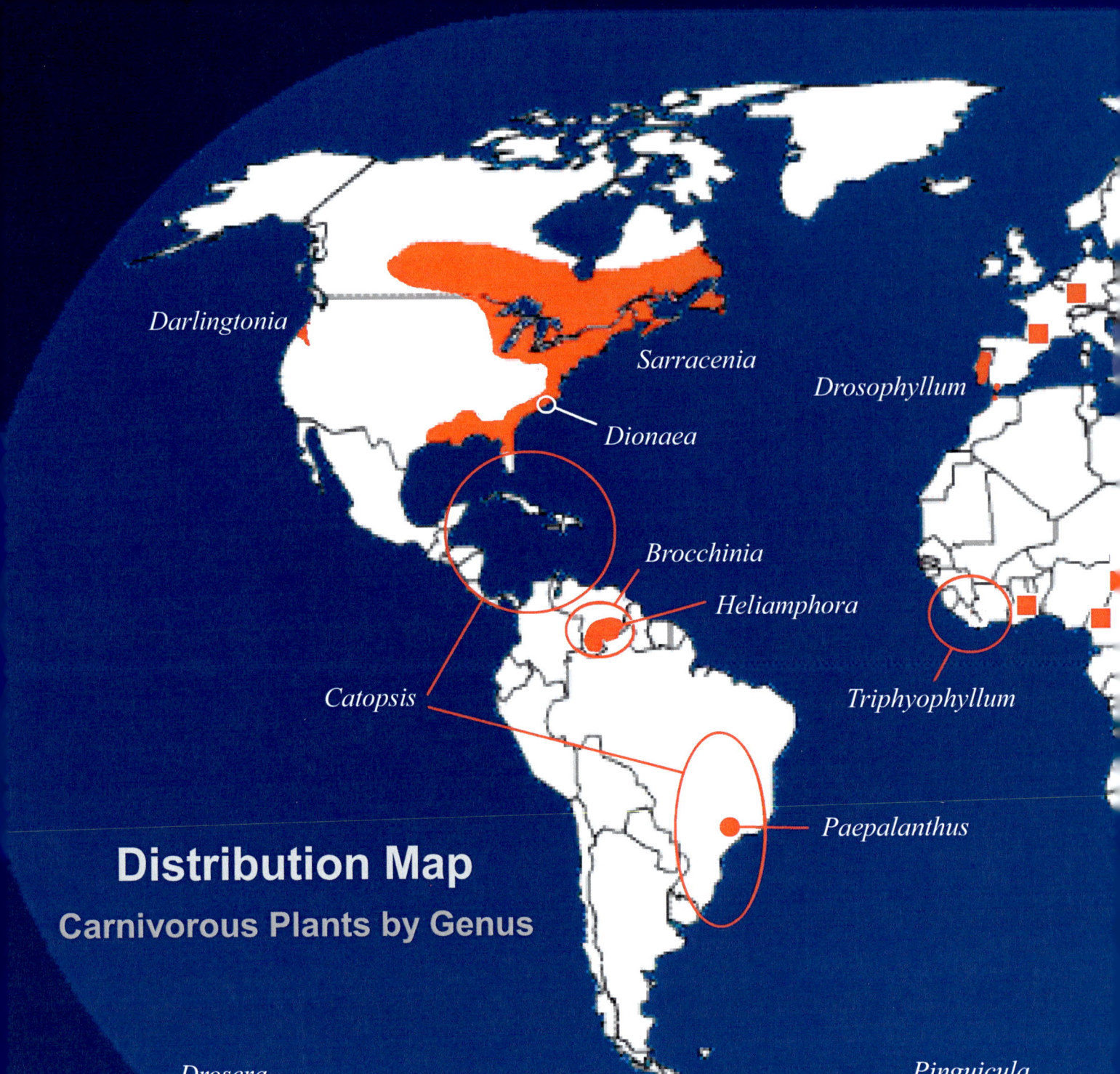

Darlingtonia

Sarracenia

Drosophyllum

Dionaea

Brocchinia

Heliamphora

Catopsis

Triphyophyllum

Paepalanthus

Distribution Map
Carnivorous Plants by Genus

Drosera

Philcoxia

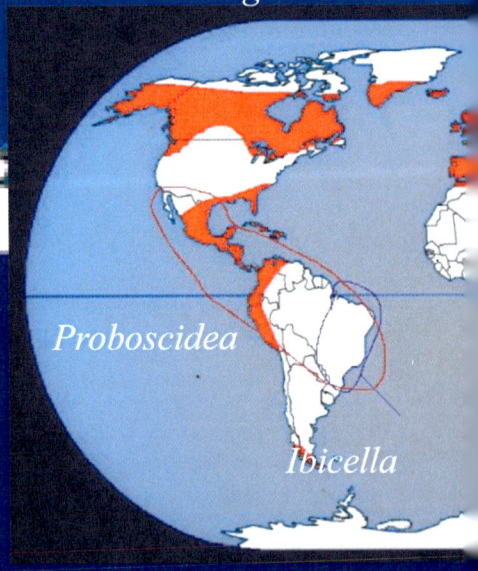

Pinguicula

Proboscidea

Ibicella

Aldrovanda

Nepenthes

Byblis

Roridula

Cephalotus

Utricularia

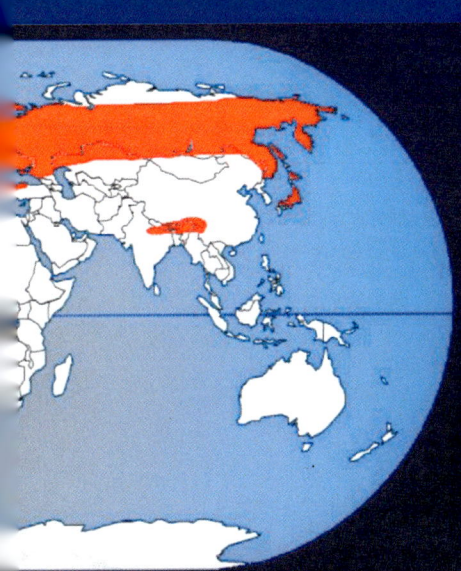

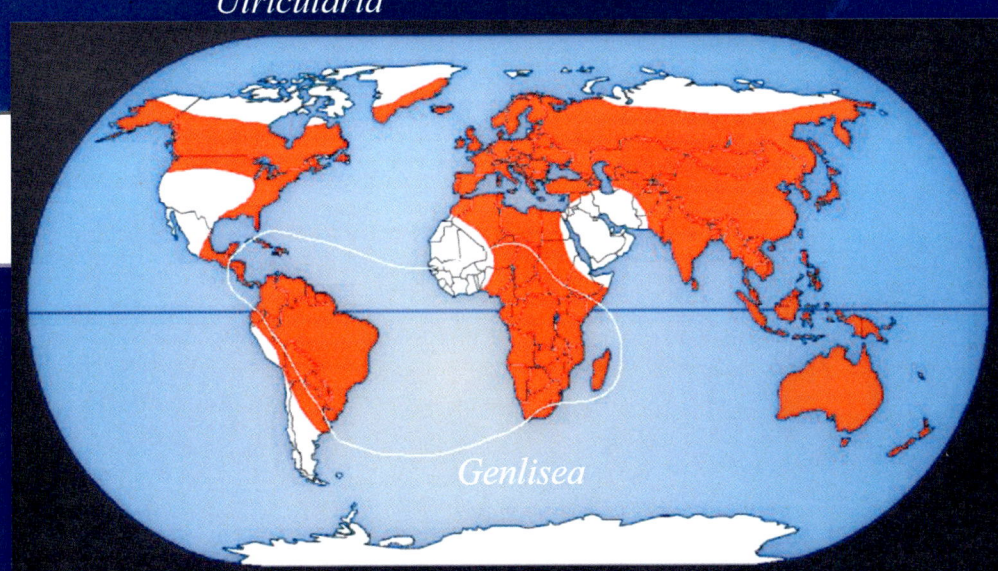

Genlisea

Glossary

abaxial underside of a leaf, facing away from the stem. cf. adaxial
actinomorphic radially symmetric flowers. cf. zygomorphic
action potential electrical pulse that travels along the cell membrane, transmitting a signal from one cell to another.
adaxial upper side of a leaf, facing toward the stem. cf. abaxial
amphiploidy process whereby a sterile hybrid organism becomes fertile due to spontaneous doubling of chromosomes.
angiosperm flowering plant. Both angiosperms and gymnosperms are seed-producing (vascular) plants. The angiosperms have the ovules and seeds enclosed in an ovary. The ancestor of angiosperms diverged from gymnosperms 200 million years ago.
anthesis period during which a flower is receptive for fertilization.
arachnid joint-legged creatures such as spiders, scorpions, ticks and mites.
arthropod the largest phylum of animals (including insects, arachnids, crustaceans, etc) characterized by their segmented body and chitinous exoskeleton.
chasmogamous flowers that open normally to allow for pollination. cf. cleistogamous
chitin main component of the exoskeletons of arthropods, such as insects.
chlorophyll green pigment found in plants that absorbs light and is critical in photosynthesis.
cleistogamous flowers that never open (but self-pollinate). cf. chasmogamous
commensal relationship in which one species is benefited while the other is unaffected.
crustacean large group of arthropods, mostly aquatic, such as crabs, lobsters, shrimp.
cuticle waxy covering of the epidermal cells of leaves to make the leaf surface impermeable (*adj.* cuticular).
decumbent leaf reclining on the ground.
digestion process whereby complex molecules are broken down into smaller structures for absorption.
dioecious plant having male and female unisexual flowers on separate plants, a la *Nepenthes*. cf. monoecious
Diptera insect order characterized by the presence of a single pair of wings, including true flies, mosquitoes, gnats, midges.
endemic native only to a certain region. cf. indigenous
ensiform having a shape of a sword.
entire having a smooth edge without teeth or lobes.
enzyme protein that accelerates a specific chemical reaction without altering itself.
epiascidiate type of leaf whose upper surface curls around and fuses to form a tube.
epidermis outermost layer of cells covering a leaf surface. cf. mesophyll
epiphyte plant that grows upon another plant (tree) without harming the host.
gemma bud-like structure formed in the rosette center, as in pygmy sundews, that develops into a new plant (*pl.* gemmae).
herbaceous plants with a non-woody stem, dies back every year. A herb.
hibernaculum protective bud made of small, tightly packed leaves that tolerates cold and desiccation (*pl.* hibernacula).
homology similarity in characters due to common ancestry, same organ in different species.
indigenous native. cf. endemic
inflorescence cluster of flowers on a stalk.
inquiline animal that uses another species for housing without hurting the host, à la microorganisms in the pitcher plant pool.
invertebrate animal without backbones. cf. vertebrate
lamina leaf blade. cf. petiole
mesophyll cells in the leaf interior that perform photosynthesis. cf. epidermis
monoecious plant having both male and female organs, either in the same flower or in different unisexual flowers. cf. dioecious
nastic movement toward a predetermined direction. cf. tropistic
ontogeny developmental change of an organism.
pedicel short stem holding each flower in an inflorescence. cf. peduncle
peduncle main stalk of an inflorescence. cf. pedicel
petiole stalk of a leaf leading to the leaf blade. cf. lamina
phyllodium flattened leaf without a pitcher tube (*pl.* phyllodia).
phylogeny evolutionary history of a species.
prostrate leaf lying flat.
protandrous flowers in which anthers release pollen before the stigma becomes receptive. cf. protogynous
protogynous flowers in which the stigma becomes receptive before pollen is shed from the anthers. cf. protandrous
protozoa single-celled, the most abundant animals in the world in number and in biomass, play a vital role in controlling bacteria.
raceme unbranching inflorescence, with each flower attached by a pedicel, the oldest flower toward the base (*adj.* racemose).
rhizome horizontal stem of a plant (usually underground).
rotifer microscopic multi-celled animal (less than 1000 cells), possessing a hair-like cilia at the body front for locomotion.
stipule small, leaf-like appendage at the base of the leaf petiole.
stolon horizontal shoot (often underground) that forms a new plant at the tip, also called a runner.
sympatry occurring in the same geographic area.
taxonomy discipline of biology to classify organisms.
tropistic movement in the direction having a correlation to the origin of stimuli. cf. nastic
vertebrate animal with backbones, such as mammals, birds, reptiles, amphibians and fish. cf. invertebrate
zygomorphic bilaterally symmetric flowers. If cut vertically, the two halves are mirror images of each other. cf. actinomorphic

Bibliography

A

Achterberg, C. van. **A Study about the Arthropoda Caught by *Drosera* Species**. *Entomologische Berichten*. 33: 137-140. 1. VII. **1973**.

Adamec, Lubomir. **Zero Water Flows in the Carnivorous Genus *Genlisea***. *Carnivorous Plant Newsletter* 32: 46-48. **2003**.

Adamec, Lubomir. **Ecophysiological Investigation on *Drosophyllum lusitanicum*: Why Doesn't the Plant Dry Out?** *Carnivorous Plant Newsletter* 38: 71-74. **2009**.

Adamec, Lubomir. **News in Ecophysiological Research on Aquatic *Utricularia* Traps**. *Carnivorous Plant Newsletter* 41: 92-104. **2012**.

Adams, Richard M., II. and Clark Barton. **The Fresh Eaters**. *Garden Journal (NYBG)* 26(5): 154-157. **1976**.

Adams, Richard M., II and George W. Smith. **An S.E.M. Survey of the Five Carnivorous Pitcher Plant Genera**. *American Journal of Botany* 64(3): 265-272. **1977**.

Addicott, John F. **Predation and Prey Community Structure: An Experimental Study of the Effect of Mosquito Larvae on the Protozoan Communities of Pitcher Plants**. *Ecology* 55: 475-492. **1974**.

Allen, Ruth McVaugh. **A Study of *Utricularia olivacea***. *Bartonia* 29: 1-2. **1957-58**.

Amagase, Shizuko. **Digestive Enzymes in Insectivorous Plants. III. Acid Proteases in the Genus *Nepenthes* and *Drosera peltata***. *Journal of Biochemistry* 72: 73-81. **1972**.

Amagase, Shizuko, Mayumi Mori and Shigeru Nakayama. **Digestive Enzymes in Insectivorous Plants. IV. Enzymatic Digestion of Insects by *Nepenthes* Secretion and *Drosera peltata* Extract: Proteolytic and Chtinolytic Activities**. *Journal of Biochemistry* 72: 765-767. **1972**.

Anderson, C. L., K. Bremer and E. M. Friis. **Dating Phylogenetically Basal Eudicots Using rbcL Sequences and Multiple Fossil Reference Points**. *American Journal of Botany* 92(10): 1737–1748. **2005**.

Anderson, Loran C. **Stalking the Pygmy Bladderwort, *Utricularia olivacea* (Lentibulariaceae)**. *Carnivorous Plant Newsletter* 29: 72-74. **2000**.

Angiosperm Phylogeny Group. **An Update of the Angiosperm Phylogeny Group Classification for the Orders and Families of Flowering Plants: APG II**. *Botanical Journal of the Linnean Society* 141(4): 399-436. **2003**.

Angiosperm Phylogeny Group. **An Update of the Angiosperm Phylogeny Group Classification for the Orders and Families of Flowering Plants: APG III**. *Botanical Journal of the Linnean Society* 161(2): 105-121. **2009**.

Angiosperm Phylogeny Group. **An Update of the Angiosperm Phylogeny Group Classification for the Orders and Families of Flowering Plants: APG IV**. *Botanical Journal of the Linnean Society* 181(1): 1-20. **2016**.

B

Bailey, Tim S. *Miraculum Naturae - Venus's Flytrap*. Trafford Publishing. **2008**.

Baldwin, Mike. *Drosera anglica* in Alaska. *Carnivorous Plant Newsletter* 42: 30-31. **2013**.

Barnhart, John Hendley. **Segregation of Genera in Lentibulariaceae**. *Memoirs of the New York Botanical Garden* 6: 39-64. **1916**.

Barnhart, John Hendley. *Pinguicula caerulea*. *Addisonia* 18: 21-22. **1933**.

Barnhart, John Hendley. *Pinguicula lutea*. *Addisonia* 18: 23-24. **1933**.

Barron, G. L. **Nematophagous Fungi: A New *Arthrobotrys* with Nonseptate Conidia**. *Canadian Journal of Botany* 57: 1371-1373. **1979**.

Barthlott, Wilhelm, Stefan Porembski, Rüdiger Seine and Inge Theisen. *The Curious World of Carnivorous Plants*. Portland, Oregon: Timber Press. **2007**.

Beal, W. J. **Carnivorous Plants**. Proceedings of the American Association for the Advancement of Science 1875B: 251-253. **1875**.

Bell, C. Ritchie. **A Cytotaxonomic Study of the Sarraceniaceae of North America**. *Journal of the Mitchell Society* 65: 137-166. **1949**.

Bell, C. Ritchie. **Natural Hybrids in the Genus *Sarracenia***. *Journal of the Mitchell Society* 68: 55-80. **1952**.

Bell, C. Ritchie. *Sarracenia leucophylla* Rafinesque. *Journal of the Mitchell Society* 70: 57-60. **1954**.

Bell, C. Ritchie and Frederick W. Case. **Natural Hybrids in the Genus *Sarracenia*. II. Current Notes on Distribution**. *Journal of the Mitchell Society* 72: 142-152. **1956**.

Benolken, R. M. and S. L. Jacobson. **Response Properties of a Sensory Hair Excised from Venus's Flytrap**. *The Journal of General Physiology* 56: 64-82. **1970**.

Bernon, Gary L. **Paper Wasp Nest in Pitcher Plant, *Sarracenia purpurea* L.** *Entomological News* 80(6): 148. **1969**.

Bradshaw, William E. **Geography of Photoperiodic Response in Diapausing Mosquito**. *Nature* 262: 384-385. **1976**.

Breckpot, Christian. *Aldrovanda vesiculosa*: **Description, Distribution, Ecology and Cultivation**. *Carnivorous Plant Newsletter* 26: 73-82. **1997**.

Bringmann, G., Schlauer, J, Wolf, K., Rischer, H., Buschboom, U., Kreiner, A., Thiele, F., Duscheck, M., & L. Ake Assi. **Cultivation of *Triphyophyllum peltatum* (Dioncophyllaceae), the part-time carnivorous plant**. *Carnivorous Plant Newsletter* 28: 7-13. **1999**.

Bringmann, G., Wenzel, M., Bringmann, H., Ake Assi, L. Hass, F., & J. Schlauer. **Uptake of the amino acid alanine by digestive leaves: proof of the carnivory in the tropical liana** *Triphyophyllum peltatum* **(Dioncophyllaceae)**. *Carnivorous Plant Newsletter* 30: 15-21. **2001**.

Bringmann, G., Rischer, H., Schlauer, J., & K. Wolf. **The tropical liana** *Triphyophyllum peltatum* **(Dioncophyllaceae): formation of carnivorous organs is only a facultative prerequisite for shoot elongation**. *Carnivorous Plant Newsletter* 31: 44-52. **2002**.

Brittnacher, John. *Drosera* x *hybrida* **Rest in Peace**. *Carnivorous Plant Newsletter* 40: 112-121. **2011**.

Burgess, L. and J. G. Rempel. **Collection of the Pitcherplant Mosquito,** *Wyeomyia smithii* **(Diptera: Culicidae), from Saskatchewan**. *The Canadian Entomologist* 103: 886-887. **1971**.

C

Cameron, Kenneth M., Kenneth J. Wurdack and Richard W. Jobson. **Molecular Evidence for the Common Origin of Snap-Traps among Carnivorous Plants**. *American Journal of Botany* 89(9): 1503-1509. **2002**.

Carlquist, Sherwin. **Wood Anatomy of Roridulaceae: Ecological and Phylogenetic Implications**. *American Journal of Botany* 63(7): 1003-1008. **1976**.

Case, Frederic W., Jr. **Some Michigan Records for** *Sarracenia purpurea* **Forma** *Heterophylla*. *Rhodora* 58: 203-207. **1956**.

Case, Frederic W. and Roberta B. Case. **The** *Sarracenia rubra* **Complex**. *Rhodora* 78: 270-325. **1976**.

Case, Frederic W., Jr. and Roberta B. Case. *Sarracenia alabamensis*, **A Newly Recognized Species from Central Alabama**. *Rhodora* 76: 650-665. **1974**.

Casper, S. Jost. **On** *Pinguicula macroceras* **Link in North America**. *Rhodora* 64: 212-221. **1962**.

Catalani, Michael. **A Field Study of** *Sarracenia oreophila*. *Carnivorous Plant Newsletter* 33: 6-12. **2004**.

Ceska, A. and M. A. M. Bell. *Utricularia* **(Lentibulariaceae) in the Pacific Northwest**. *Madrono* 22: 74-84. **1973**.

Chamovitz, Daniel. *What a Plant Knows*. New York: Scientific American / Farrar, Straus and Giroux. **2012**.

Chan, Michael M., Mallory M. Chan and Edward D. Chan. **What Is the Evidence for Medicinal Value of Carnivorous Plants?** *Carnivorous Plant Newsletter* 36: 83-86. **2007**.

Cheek, Martin and Malcolm Young. **The Limonium Pregrinum of Carolus Clusius**. *Carnivorous Plant Newsletter* 23: 95-98. **1994**.

Cheek, Martin. **The Correct Names for the Subspecies of** *Sarracenia purpurea* **L.** *Carnivorous Plant Newsletter* 23: 69-73. **1994**.

Christensen, Norman L. **The Role of Carnivory in** *Sarracenia flava* **L. with Regard to Specific Nutrient Deficiencies**. *Journal of the Mitchell Society* 92: 144-147. **1977**.

Christophe, Alain. **A South African Insectivorous Plant Trip**. *Carnivorous Plant Newsletter* 10: 96-101. **1981**.

Clancy, Finbarr G. and Michael D. Coffey. **Acid Phosphatase and Protease Release by the Insectivorous Plant** *Drosera rotundifolia*. *Canadian Journal of Botany* 55: 480-488. **1977**.

Claudi-Magnussen, Glenn. **An Introduction to** *Genlisea*. *Carnivorous Plant Newsletter* 11: 13-16. **1982**.

Cody, William and Stephen S. Talbot. **The Pitcher Plant** *Sarracenia purpurea* **L. in the Northwestern Part of its Range**. *The Canadian Field-Naturalist* 87: 318-320. **1973**.

Conran, John G., Gunta Jaudzems and Neil D. Hallam. **Droseraceae Gland and Germination Patterns Revisited: Support for Recent Molecular Phylogenetic Studies**. *Carnivorous Plant Newsletter* 36: 14-20. **2007**.

Cronquist, Arthur. *The Evolution and Classification of Flowering Plants*. Houghton Mifflin. **1968**.

Cross, Adam. *Aldrovanda — The Waterwheel Plant*. Poole, Dorset, England: Redfern Natural History Productions. **2012**.

Cruise, James E. and Paul M. Catling. **The Sundews (** *Drosera* **spp.) in Ontario**. *The Ontario Field Biologist* 28: 1-6. **1974**.

D

D'Amato, Peter. *The Savage Garden*. Berkeley, California: Ten Speed Press. **1998**.

Darling, Thomas, Jr. **Insectivorous Plants in the Poconos**. *Castanea* 29: 126-128. **1964**.

Darling, Thomas, Jr. and Stanwyn G. Shetler. *Sarracenia* x *catesbaei* **Elliott (pro sp.) in the Pocono Mountains of Pennsylvania**. *Castanea* 37(2): 133-137. **1972**.

Darnowski, Douglas W. and Sarah Fritz. **Prey Preference in** *Genlisea* **Small Crustaceans, Not Protozoa**. *Carnivorous Plant Newsletter* 39: 114-116. **2010**.

Darwin, Charles. *Insectivorous Plants*. London: John Murray. **1875**.

Darwin, Charles. *Insectivorous Plants*. 2nd ed., revised by Francis Darwin. London: John Murray, Albemarle Street. **1893**.

DeBuhr, Larry. **The Distribution of** *Darlingtonia californica*. Carnivorous Plant Newsletter 3: 24-26. **1974**.

DeBuhr, Larry E. **Phylogenetic Relationships of the Sarraceniaceae**. *Taxon* 24(2/3): 297-306. **1975**.

DeBuhr, Larry E. **Wood Anatomy of the Sarraceniaceae; Ecological and Evolutionary Implications**. *Plant Systematics and Evolution* 128: 59-169. **1977**.

Degreef, John. **The Electrochemical Mechanism of Trap Closure in** *Dionaea muscipula*. *Carnivorous Plant Newsletter* 17: 80-83,94. **1988**.

Degreef, John. **The Evolution of** *Aldrovanda* **and** *Dionaea* **Traps**. *Carnivorous Plant Newsletter* 17: 119-124. **1988**.

Degreef, John. **Early History of** *Drosera* **and** *Drosophyllum*. *Carnivorous Plant Newsletter* 18: 86-89. **1989**.

Degreef, John. **Evolutionary Patterns in** *Drosera*. *Carnivorous Plant Newsletter* 19: 11-16. **1990**.

Degreef, John D. **More on the Evolution of *Drosera***. *Carnivorous Plant Newsletter* 19: 92. **1990**.

Degreef, John D. **The Origin of the Genus *Byblis***. *Carnivorous Plant Newsletter* 19: 93-95. **1990**.

Degreef, John D. ***Cephalotus follicularis*: History and Evolution**. *Carnivorous Plant Newsletter* 19: 95103. **1990**.

Degreef, John D. **Fossil *Aldrovanda***. *Carnivorous Plant Newsletter* 26: 93-97. **1997**.

Devienne, Karina F., Maria Stella G. Raddi, Eliana A. Varanda and Wagner Vilegas. **In Vitro Cytotoxicity of Some Natural and Semi-Synthetic Isocoumarins from *Paepalanthus bromelioides***. *Z. Naturforsch*. 57c: 85-88. **2002**.

Dunsterville, G. C. K. ***Zygosepalum tatei* in One of Its Natural Haunts**. *Orchid Digest* 35: 239-242. **1971**.

E

Elder, Christine Leigh. **Reproductive Biology of *Darlingtonia californica***. *Fremontia. The Journal of the California Native Plant Society*. October **1994**.

Ellison, Aaron M., Hannah L. Buckley, Thomas E. Miller and Nicholas J. Gotelli. **Morphological Variation in *Sarracenia purpurea* (Sarraceniaceae): Geographic, Environmental, and Taxonomic Correlates**. *American Journal of Botany* 91: 1930-1935. **2004**.

Ellison, A. M. **Nutrient Limitation and Stoichiometry of Carnivorous Plants**. *Plant Biology* 8: 740-747. **2006**.

F

Fashing, Norman James. **Arthropod Associates of the Cobra Lily (*Darlingtonia californica*)** *Virginia Journal of Science* 23(3): 92. **1981**.

Fashing, Norman James. **Biology of *Sarraceniopus darlingtoniae* (Histiostomatidae: Astigmata), An Obligatory Inhabitant of the Fluid-Filled Pitchers of *Darlingtonia californica* (Sarraceniaceae)**. *Phytophaga* XIV: 299-305. **2004**.

Figueira, J. E. C., J. Vasconcellos-Neto and P. Jolivet. **A New Protocarnivorous Plant, *Paepalanthus bromelioides* Silveira (Eriocaulaceae) from Brazil**. *Revue d'ecologie* 49(1):3-9. **1994**.

Fish, Durland and Donald W. Hall. **Succession and Stratification of Aquatic Insects Inhabiting the Leaves of the Insectivorous Pitcher Plants, *Sarracenia purpurea***. *American Midland Naturalist* 99(1): 172-183. **1978**.

Fish, Durland. **Insect-Plant Relationships of the Insectivorous Pitcher Plant *Sarracenia minor***. *The Florida Entomologist* 59(2): 199-203. **1976**.

Fleischmann, Andreas. ***Philcoxia*: A New Genus of Carnivorous Plant**. *Carnivorous Plant Newsletter* 41: 77-81. **2012**.

Fleischmann, Andreas. *Monograph of the Genus Genlisea*. Poole, Dorset, England: Redfern Natural History Productions. **2012**.

Forsyth, Adrian B. and Raleigh J. Robertson. **K Reproductive Strategy and Larval Behavior of the Pitcher Plant Sarcophagid Fly, *Blaesoxipha fletcheri***. *Canadian Journal of Zoology* 53(2): 174-179. **1975**.

Forterre, Yoel, Jan M. Skotheim, Jacques Dumais and L. Mahadevan. **How the Venus Flytrap Snaps**. *Nature* 433: 421-425. **2005**.

Franck, Daniel. H. **Early Histogenesis of the Adult Leaves of *Darlingtonia californica* (Sarraceniaceae) and Its Bearing on the Nature of Epiascidiate Foliar Appendages**. *American Journal of Botany* 62(2): 116-132. **1975**.

Franck, Daniel. H. **Comparative Morphology and Early Leaf Histogenesis of Adult and Juvenile Leaves of *Darlingtonia californica* and Their Bearing on the Concept of Heterophylly**. *Botanical Gazette* 137(1): 20-34. **1976**.

Frank, J. H. and G. F. O'Meara. **The Bromeliad *Catopsis berteroniana* Traps Terrestrial Arthropods But Harbors *Wyeomyia* Larvae (Diptera: Culicidae)**. *Florida Entomologist* 67: 418-424. **1984**.

G

Garrido, B, A. Hampe, T. Maranon and J. Arroyo. **Regional Differences in Land Use Affect Population Performance of the The Threatened Insectivorous Plant *Drosophyllum lusitanicum* (Droseraceae)**. *Diversity and Distribution* 9: 335-350. **2003**.

Gibson, Robert. **Observations on *Cephalotus* in the Wild**. *Carnivorous Plant Newsletter* 28: 30-31. **1999**.

Gibson, Robert. **Red *Aldrovanda* from Near Esperance, Western Australia**. *Carnivorous Plant Newsletter* 33: 119-121. **2004**.

Gibson, Thomas C. and Donald M. Waller. **Evolving Darwin's most wonderful plant: ecological steps to a snap-trap**. *New Phytologist* 183: 575-587. **2009**.

Gibson, Dorothy Nash. **Flora of Guatemala - Part X, Number 4 - Lentibulariaceae, Bladderwort Family**. *Fieldiana Botany* 24: 315-328. **1974**.

Gilchrist, A. J. and B. E. Juniper. **An Excitable Membrane in the Stalked Glands of *Drosera capensis* L.**. *Planta (Berl.)* 119: 143-147. **1974**.

Givnish, T. J., E. L. Burkhardt, R. E. Happel and J. D. Weintraub. **Carnivory in the Bromeliad *Brocchinia reducta*, with a Cost/Benefit Model for the General Restriction of Carnivorous Plants to Sunny, Moist, Nutrient-Poor Habitats**. *American Naturalist* 124: 479-497. **1984**.

Gleason, H. A. **Botanical results of the Tyler-Duida Expedition**. *Bulletin of the Torrey Botanical Club* 58: 277-506. **1931**.

Godfrey, R. K. and H. Larry Stripling. **A Synopsis of *Pinguicula* (Lentibulariaceae) in the Southeastern United States**. *The American Midland Naturalist* 66(2): 395-409. **1961**.

Gon, Sam, III. **The Hawaii Population of *Drosera anglica* - a Tropical Twist on a Temperate Theme**. *Carnivorous Plant Newsletter* 23: 68-69. **1994**.

Green, Sally., T. L. Green and Y. Heslop-Harrison. **Seasonal Heterophylly and Leaf Gland Features in *Triphyophyllum* (Dioncophyllaceae) A New Carnivorous Plant Genus**. *Botanical Journal of the Linnaean Society* 78: 99-116. **1979**.

H

Haber, Erich. ***Utricularia geminiscapa* at Mer Bleue and Range Extensions in Eastern Canada**. *Canadian Field-Naturalist* 93(4): 391-398. **1979**.

Haberlandt, Gottlieb. **Sinnesorgane im Pflanzenreich - V. Insectivores: *Aldrovanda vesiculosa***. *Carnivorous Plant Newsletter* 10: 73,76-79. **1981**.

Haberlandt, Gottlieb. **Sinnesorgane im Pflanzenreich - V. Insectivores: *Aldrovanda vesiculosa***. *Carnivorous Plant Newsletter* 10: 89,92-93. **1981**.

Haberlandt, Gottlieb. **Sinnesorgane im Pflanzenreich - Insectivores: *Dionaea muscipula***. *Carnivorous Plant Newsletter* 11: 9,12,21. **1982**.

Haberlandt, Gottlieb. **Sinnesorgane im Pflanzenreich - Insectivores: *Dionaea muscipula***. *Carnivorous Plant Newsletter* 11: 32-40. **1982**.

Haberlandt, Gottlieb. **Sinnesorgane im Pflanzenreich - V. Insectivores: *Drosera* and *Drosophyllum***. *Carnivorous Plant Newsletter* 11: 66-73. **1982**

Hansen, Carlo. **Note on *Drosera rotundifolia* L. in Greenland**. *Botanisk Tidsskrift (Copenhagen)* 67(4): 342-343. **1973**.

Harper, Roland M. **Botanical Explorations in Georgia During the Summer of 1901. II Noteworthy Species**. *Bulletin of the Torrey Botanical Club* 30: 319-342. **1903**.

Harper, Roland M. **Explorations in the Coastal Plain of Georgia During the Season of 1902**. *Bulletin of the Torrey Botanical Club* 31: 21-23. **1904**.

Harper, Roland M. **Phytogeographical Explorations in the Coastal Plain of Georgia In 1904**. *Bulletin of the Torrey Botanical Club* 32: 451-467. **1905**.

Harper, Roland M. **Some New or Otherwise Noteworthy Plants from the Coastal Plain of Georgia**. *Bulletin of the Torrey Botanical Club* 33: 229-245. **1906**.

Harper, Roland M. **The American Pitcher-Plants**. *Journal of the Mitchell Society* 34: 110-125. **1918**.

Hartmeyer, Irmgard and S. R. H. Hartmeyer. **Snap-Tentacles and Runway Lights: Summary of Comparative Examination of *Drosera* Tentacles**. *Carnivorous Plant Newsletter* 39: 101-113. **2010**.

Hartmeyer, Irmgard and Siegfried R. H. **Comparison of *Byblis* 'Goliath' (*B. filifolia*), *Byblis* 'David' (*B. liniflora*), and Their Putative Fertile Hybrid**. *Carnivorous Plant Newsletter* 40: 129-135. **2011**.

Hartmeyer, Siegfried. **Carnivory in *Byblis* Revisited II: The Phenomenon of Symbiosis on Insect Trapping Plants**. *Carnivorous Plant Newsletter* 27: 110-114. **1998**.

Hartmeyer, Siegfried R. H., Irmgard Hartmeyer, Tom Masselter, Robin Seidel, Thomas Speck and Simon Poppinga. **Catapults into a Deadly Trap: The Unique Prey Capture Mechanism of *Drosera Glanduligera***. *Carnivorous Plant Newsletter* 42: 4-14. **2013**.

Hasebe, Mitsuyasu. **How Carnivorous Plants Have Evolved from Non-carnivores**. *Iden (heritage). Science of Living Matter.* Volume 59. No.4: 33-37. **2005**.

Haston, E., Richardson, J.E., Stevens, P.F., Chase, M.W., Harris, D.J. **The Linear Angiosperm Phylogeny Group(LAPG) III: a linear sequence of the families in APG III**. *Botanical Journal of the Linnean Society* **161**, 128–131. **2009**.

Hellquist, C. Barre. **A White-Flowered Form of *Utricularia purpurea* from New Hampshire**. *Rhodora* 76: 19. **1974**.

Hermanova, Z., Kvacek, J. **Late Cretaceous *Palaeoaldrovanda*, Not Seeds of a Carnivorous Plant, But Eggs of an Insect**. *Journal of the National Museum (Prague), Natural History Series* 179(9): 105-118. **2010**

Heslop-Harrison, Yolande. **Scanning Electron Microscopy of Fresh Leaves of *Pinguicula***. *Science* 167: 172-174. **1970**.

Heslop-Harrison, Yolande and R. B. Knox. **A Cytochemical Study of the Leaf-Gland Enzymes of Insectivorous Plants of the Genus *Pinguicula***. *Planta (Berl.)* 96: 183-211. **1971**.

Heslop-Harrison, Yolande. **Carnivorous Plants a Century After Darwin**. *Endeavor* 35: 114-122. **1976**.

Hetch, Adolph. **The Somatic Chromosomes of *Sarracenia***. *Bulletin of the Torrey Botanical Club* 76(1): 7-9. **1949**.

Hilu, Khidir W.,Thomas Borsch, Kai Muller, Douglas E. Soltis, Pamela S. Soltis, Vincent Savolanen, Mark W. Chase, Martyn P. Powell, Lawrence A. Alice, Roger Evans, Herve Sauquet, Christoph Neinhuis, Tracy A. B. Slotta, Jens G. Rohwer, Christopher S. Cambell and Lars W. Chatrou. **Angiosperm phylogeny based on <011>matK sequence information**. *American Journal of Botany* 90: 1758-1776. **2003**.

Hobbhahn, N, H. Kuchmeister and S. Porembski. **Pollination Biology of Mass Flowering Terrestrial *Utricularia* Species (Lentibulariaceae) in the Indian Western Ghats**. *Plant Biology (Stuttg)* 8: 791-804. **2006**.

Hodick, Dieter and Andreas Sievers. **On the Mechanism of Trap Closure of Venus Flytrap (*Dionaea muscipula* Ellis)**. *Planta* 179: 32-42. **1989**.

I

Istock, Conrad A., Steven S. Wasserman and Harold Zimmer. **Ecology and Evolution of the Pitcher-Plant Mosquito: 1. Population Dynamics and Laboratory Responses to Food and Population Density**. *Evolution* 29: 296-312. **1975**.

J

Jaffe, M. J. **The Role of ATP in Mechanically Stimulated Rapid Closure of the Venus's Flytrap**. *Plant Physiology* 51: 17-18. **1973**.

Jentsch, J. **Enzymes from Carnivorous Plants (*Nepenthes*). Isolation of the Protease Nepenthacin**. *FEBS Letters* 21: 273-276. **1972**.

Jobson, Richard W., Rasmus Nielsen, Liisa Laakkonen, Marten Wikstrom and Victor A. Albert. **Adaptive Evolution of Cytochrome C Oxidase: Infrastructure for a Carnivorous Plant Radiation**. *PNAS* 101(52): 18064-18068. **2004**.

Joel, Daniel M. **Mimicry in Carnivorous Pitcher Plants - Fact or Legend?** *Carnivorous Plant Newsletter* 18: 12-14. **1989**.

Jolivet, Pierre. *Interrelationships between Insects and Plants*. CRC Press LLC. **1998**.

Jones, Frank Morton. **Pitcher-Plant Insects**. *Entomological News* 15: 14-17. **1904**.

Jones, Frank Morton. **Pitcher-Plant Insects-II**. *Entomological News* 18: 413-420. **1907**.

Jones, Frank Morton. **Pitcher-Plant Insects-III**. *Entomological News* 19: 150-156. **1908**.

Jones, Frank Morton. **Another Pitcher-Plant Insect (Diptera, Sciarinae)**. *Entomological News* 31: 91-94. **1920**.

Jones, Frank Morton. **Pitcher Plants and Their Moths**. *Natural History* 21: 296-316. **1922**.

Joyeux, M., Vincent, O., and Marmottant, P. **Mechanical Model of the Utrafast Underwater Trap of *Utricularia***. *Physical Review*. E83. 021911. **2011**.

Judd, W. W. **Studies of the Byron Bog in Southwestern Ontario X. Inquilines and Victims of the Pitcher-Plant, *Sarracenia purpurea* L**. *The Canadian Entomologist* 91: 171-180. **1959**.

Juniper, B. E., R. J. Robins, and D. M. Joel. *The Carnivorous Plants*. San Diego: Academic Press Limited. **1989**.

K

Komiya, Sadashi, and Kiyoshi Shimizu. *Insectivorous Plants*. Tokyo: New Science. **1978**.

Komiya, Sadashi. *The Insectivorous Plants*. Tokyo: Japan Insectivorous Plant Society Publication. **1994**.

Kondo, Katsuhiko. **Chromesome Number of Carnivorous Plants**. *Bulletin of the Torrey Botanical Club* 96(3): 322-328. **1969**.

Kondo, Katsuhiko. **A New Species of *Nepenthes* from the Philippines**. *Bulletin of the Torrey Botanical Club* 96(6): 653-655. **1969**.

Kondo, Katsuhiko. **A Comparison of Variability in *Utricularia cornuta* and *U. juncea***. *American Journal of Botany* 59 (1): 23-37. **1972**.

Kondo, Katsuhiko. **The Karyotypes of the Species of *Byblis***. *Bulletin of the Torrey Botanical Club* 100: 367-369. **1973**.

Kondo, Katsuhiko. **The Sunshine Pitchers**. *Garden Journal (NYBG)* 24(1): 14-15. **1974**.

Kondo, Katsuhiko, Michiharu Segawa and Kunito Nehira. **Anatomical Studies on Seeds and Seedlings of Some *Utricularia* (Lentibulariaceae)**. *Brittonia* 30: 89-95. **1978**.

Korolas, Jim. **Cultivating *Drosera linearis* (Goldie)**. *Carnivorous Plant Newsletter* 11: 19-20,27. **1982**

Krajina, V. J. **Sarraceniaceae, A New Family for British Columbia**. *Syesis* 1: 121-124. **1968**.

Kurata, Shigeo. **Biology of Nepenthes**. *Iden (Heritage)* 26(10): 43-51. **1972**.

L

Lamb, Randy. *Pinguicula villosa* **The Northern Butterwort**. *Carnivorous Plant Newsletter* 20: 73-77. **1991**.

Lavarack, P. S. **The Northern Rainbow Plant - *Byblis liniflora***. *Carnivorous Plant Newsletter* 10: 102-103. **1981**.

Legendre, Laurent and Thomas Cieslak. **Pinguicula vulgaris L. in the Champagne State of France: Life in an Alkaline Bog**. *Carnivorous Plant Newsletter* 36: 104-113. **2007**.

Li, Hongqi. **Observation of Reproductive Organs of Sarraceniaceae with SEM LV Model**. *Carnivorous Plant Newsletter* 39: 56-61. **2010**.

Lichtner, F. T. and R. M. Spanswick. **Ion Relations in Dionaea**. *Plant Physiology* 59, Suppl., 84. **1977**.

Lloyd, Francis E. *The Carnivorous Plants*. Waltham, Mass.: Chronica Botanica Company. **1942**.

Lowrie, Allen. *Carnivorous Plants of Australia*. Vol.1. University of Western Australia Press. **1987**.

Lowrie, Allen. *Carnivorous Plants of Australia*. Vol.2. University of Western Australia Press. **1989**.

Lowrie, Allen. *Carnivorous Plants of Australia*. Vol.3. University of Western Australia Press. **1998**.

M

MacFarlane, J. M. **Observations on *Sarracenia***. *Journal of Botany* 45: 1-7. **1907**.

Maguire, Bassett, Richard S. Cowan, John J. Wurdack and Collaborators. **The Botany of the Guayana Highland**. *Memoirs of the New York Botanical Garden* 8(2): 87-96. **1953**.

Maguire, Bassett. **On the Flora of the Guayana Highland**. *Biotropica* 2(2): 85-100. **1970**.

Maguire, Bassett. Wurdack and Collaborators. **The Botany of the Guayana Highland**. *Memoirs of the New York Botanical Garden* 29(2): 87-96. **1978**.

Mameli, E. **Ricerche anatomiche, fisiologiche e biologiche sulla *Martynia lutea* Lindl. (Anatomical, physiological and biological research on *Martynia lutea*)**. *Atti dell' Universita di Pavia*, Serie 2, 16, 137-188. **1916**.

Mandossian, Adrienne J. **Plant Associates of *Sarracenia purpurea* (Pitcher Plant) in Acid and Alkaline Habitats**. *The Michigan Botanist* 4: 107-114. **1965**.

Mandossian, Adrienne J. **Variations in the Leaf of *Sarracenia purpurea* (Pitcher Plant)**. *The Michigan Botanist* 5: 26-35. **1966**.

Mandossian, Adrienne J. **Germination of Seeds in *Sarracenia purpurea* (Pitcher Plant)**. *The Michigan Botanist* 5: 67-79. **1966**.

Markgraf, F. **Uber Laubblatt-Homologien und Verwandtschaftliche Zusammenhange**. *Planta, Bd.* 46: 414-446. **1955**.

McDaniel, Sidney. **The Genus *Sarracenia* (Sarraceniaceae)**. *Bulletin of Tall Timbers Research Station*. No. 9. **1971**.

McPherson, Stewart. *Pitcher Plants of the Americas*. Blacksburg, Virginia: The McDonald & Woodward Publishing Company. **2007**.

McPherson, Stewart. *Glistening Carnivores — The Sticky-Leaved Insect-Eating Plants*. Poole, Dorset, England: Redfern Natural History Productions. **2009**.

McPherson, Stewart. *Pitcher Plants of the Old World*. Vol. 1 & 2. Poole, Dorset, England: Redfern Natural History Productions. **2009**.

McPherson, Stewart. *Carnivorous Plants and Their Habitats*. Vol. 1 & 2. Poole, Dorset, England: Redfern Natural History Productions. **2010**.

McPherson, Stewart and Donald Schnell. *Sarraceniaceae of North America*. Poole, Dorset, England: Redfern Natural History Productions. **2011**.

McPherson, Stewart, Andreas Wistuba, Andreas Fleischmann and Joachim Nerz. *Sarraceniaceae of South America*. Poole, Dorset, England: Redfern Natural History Productions. **2011**.

Meindl, George A. **Pollination Biology of *Darlingtonia californica*, the California Pitcher Plant**. Master's Thesis, Humboldt State University. **2009**.

Mellichamp, Larry. **Botanical History of CP II: *Darlingtonia***. *Carnivorous Plant Newsletter* 7: 82-85. **1978**.

Mellichamp, Larry. **Field Studies on CP at UMBS**. *Carnivorous Plant Newsletter* 11: 10-11,13. **1982**.

Mellichamp, T. Lawrence. **Cobras of the Pacific Northwest**. *Natural History* 92: 46-51. **1983**.

Meyers-Rice, Barry A. **An Anthocyanin-Free Variant of *Darlingtonia californica*: Newly Discovered and Already Imperiled**. *Carnivorous Plant Newsletter* 26: 129-132. **1997**.

Meyers-Rice, Barry A. *Darlingtonia californica* **"Othello."** *Carnivorous Plant Newsletter* 27: 40-42. **1998**.

Miles, D. Howard, Udom Kokpol, Leon H. Zalkowi, Steven J. Steindeli and James B. Naborsi. **Tumor Inhibitors I: Preliminary Investigation of Antitumor Activity of *Sarracenia flava***. *Journal of Pharmaceutical Sciences* 63: 613-615. **1974**.

Miles, D. Howard, Udom Kokpol, Naresh V. Mody and Paul A. Hedin. **Volatiles in *Sarracenia flava***. *Photochemistry* 14: 845-846. **1975**.

Mohr, Charles. **Notes on Some Undescribed and Little Known Plants of the Alabama Flora**. *Bulletin of the Torrey Botanical Club* 24: 19-32. **1897**.

Moldenke, Harold N. **Notes on New and Noteworthy Plants. LX**. *Phytologia* 26(4): 224-226. **1973**.

Molis, Arne, Gabriele Patten and Manfred Weidner. **The Time Memory of the Venus Flytrap (Dionaea muscipula Ellis)**. *Carnivorous Plant Newsletter* 35: 108-118. **2006**.

Mozingo, Hugh N. **Venus Flytrap Observations by Scanning Electron Microscopy**. *American Journal of Botany* 57(5): 593-598. **1970**.

Murry, Robert E., Jr. and Lowell E. Urbatsch. **Preliminary Reports on the Flora of Louisiana. III. The Families Droseraceae and Sarraceniaceae**. *Castanea* 44: 24-27. **1979**.

Murza, Gillian L. and Arthur R. Davis. **Comparative Flower Structure of Three Species of Sundew (*Drosera anglica, Drosera linearis, and Drosera rotundifolia*) in Relation to Breeding System**. *Canadian Journal of Botany* 81(11): 1129-1142. **2003**.

Murza, Gillian L. and Arthur R. Davis. **Flowering Phenology and Reproductive Biology of *Drosera anglica* (Droseraceae)**. *Botanical Journal of Linnean Society* 147(4): 417-426. **2005**.

Murza, Gillian L., Joanne R. Heaver and Arthur R. Davis. **Minor Pollinator-Prey Conflict in the Carnivorous Plant, *Drosera anglica***. *Plant Ecology* 184: 43-52. **2006**.

N

Naeem, Shahid. **Resource Heterogeneity Fosters Coexistence of a Mite and a Midge in Pitcher Plants**. *Ecological Monographs* 58(3): 215-227. **1988**.

Neid, Stephanie L. *Utricularia minor* **L. (Lesser Bladderwort) - A Technical Conservation Assessment**. USDA Forest Service, Rocky Mountain Region, Species Conservation Project. May 15, **2006**.

Nelson, E. Charles and Daniel L. McKinley. *Aphrodite's Mousetrap. A Biology of Venus's Flytrap with Facsimiles of John Ellis's Original Pamphlet and Manuscripts*. Boethius Press. 1990. A facsimile published by Boethius Press in association with the Linnean Society of London and with the assistance of the Bentham-Moxon Trust, Royal Botanic Gardens, Kew. **1990**.

Nerz, Joachim. *Heliamphora elongata* **(Sarraceniaceae), a New Species from Ilu-Tepui**. *Carnivorous Plant Newsletter* 33: 111-116. **2004**.

Nesbitt, Herbert H. J. **A New Mite, *Zwickia gibsoni* n.sp., Fam. Anoetidae, from the Pitchers of *Sarracenia purpurea* L**. *The Canadian Entomologist* 86: 193-197. **1954**.

Neyland, Ray and Mark Merchant. **Systematic Relationships of Sarraceniaceae Inferred from Nuclear Ribosomal DNA Sequences**. *Madrono* 53(3): 223-232. **2006**.

Nichols, M. Louse. **The Development of the Pollen of *Sarracenia*.** *Botanical Gazette* 45: 31-37. **1908**.

Nielsen, David W. **Arthropod Communities Associated with *Darlingtonia californica*.** *Annals of the Entomological Society of America* 83(2): 189-200. **1990**.

Nolan, Garry. **On the Foraging Strategies of Carnivorous Plants: II. Biological Stimulus verses Mechanical Stimulus in the Fast-Moving Periphery Tentacles of the Species *Drosera burmanni*.** *Carnivorous Plant Newsletter* 7: 79-81. **1978**.

Nyoka, Susan E. and Carol Ferguson. **Pollinators of *Darlingtonia californica* Torr., the California Pitcher Plant.** *Natural Areas Journal* 19(4): 386-391. **1999**.

Nyoka, Susan E. **The Spider and the Fly: A Proposed Pollination Scenario for *Darlingtonia californica*, the California Pitcher Plant.** *Bulletin of the Native Plant Society of Oregon* 68-69. **2000**.

O

O'Neal, Wendy. **A Preliminary Report on the Pollination of a *Sarracenia purpurea* in a Forest-Swale Ecotone.** *Carnivorous Plant Newsletter* 12: 60-61,74. **1983**.

Opel, Matthew R. ***Roridula*, A Carnivorous Shrub from South Africa.** *Carnivorous Plant Newsletter* 34: 106-110. **2005**.

Owen, T. Page, Jr. and Kristen A. Lennon. **Structure and Development of the Pitchers from the Carnivorous Plant *Nepenthes alata* (Nepenthaceae).** *American Journal of Botany* 86(10): 1382-1390. **1999**.

P

Pagano, M. C. and M. R. Scotti. **A Survey of the Arbuscular Mycorrhiza Occurrence in *Paepalanthus bromelioides* and *Bulbostylis* sp. in Rupestrian Fields, Brazil.** *Micologia Aplicada International* 21(1): 1-10. **2009**.

Pietropaolo, James and Patricia A. *The World of Carnivorous Plants*. Shortsville, New York: R. J. Stoneridge. **1974**.

Plummer, Gayther L. and Thomas H. Jackson. **Bacteria Activities within the Sarcophagus of the Insectivorous Plant, *Sarracenia flava*.** *American Midland Naturalist* 69(2): 462-469. **1963**.

Plummer, Gayther L. **Soils of the Pitcher Plant Habitats in the Georgia Coastal Plain.** *Ecology* 44: 727-734. **1963**.

Plummer, Gayther L. and John B. Kethley. **Foliar Absorption of Amino Acids, Peptides, and Other Nutrients by the Pitcher Plant, *Sarracenia flava*.** *Botanical Gazette* 125(4): 245-260. **1964**.

R

Ragetli, H. W. J., M. Weintraub and Esther Lo. **Characteristics of *Drosera* Tentacles. I. Anatomical and Cytological Detail.** *Canadian Journal of Botany* 50: 159-168. **1972**.

Raju, M. V. S. Development of **Floral Organs in the Sites of Leaf Primordia in *Pinguicula vulgaris*.** *American Journal of Botany* 56(5): 507-514. **1969**.

Rao, A. N. and E. T. Ong. **Germination of Compound Pollen Grains.** *Grana* 12: 113-120. **1972**.

Reifenrath, Kerstin, Inge Theisen, Jan Schnitzler, Stefan Porembski and Wilhelm Barthlott. **Trap Architecture in Carnivorous *Utricularia* (Lentibulariaceae).** *Flora - Morphology, Distribution, Functional Ecology of Plants* 201(8): 597-605. **2006**.

Reinert, Grady W. and R. K. Godfrey. **Reappraisal of *Utricularia inflata* and *U. radiata* (Lentibulariaceae).** *American Journal of Botany* 49: 213-220. **1962**.

Rice, Barry. **Testing the Appetites of *Ibicella* and *Drosophyllum*.** *Carnivorous Plant Newsletter* 28: 40-43. **1999**.

Rice, Barry A. *Growing Carnivorous Plants*. Portland, Oregon: Timber Press. **2006**.

Rice, Barry. **Reassessing Commensal-Enabled Carnivory in *Proboscidea* and *Ibicella*.** *Carnivorous Plant Newsletter* 37: 15-19. **2008**.

Rice, Barry, Arthur Yin and Gina E. Morimoto. **Observations of Isolated *Pinguicula* Populations in the Western USA.** *Carnivorous Plant Newsletter* 37: 100-109. **2008**.

Rice, Barry. **Flower Studies Do Not Support Subspecies Within *Pinguicula macroceras*.** *Carnivorous Plant Newsletter* 40: 44-49. **2011**.

Rice, Barry. **Tuberous Organs in *Utricularia*, and New Observations of Sub-Tuberous Stolons on *Utricularia radiata* Small.** *Carnivorous Plant Newsletter* 40: 88-91. **2011**.

Rice, Barry. **The Thread-Leaf Sundews *Drosera filiformis* and *Drosera tracyi*.** *Carnivorous Plant Newsletter* 40: 4-16. **2011**.

Richards, Jennifer H. **Bladder Function in *Utricularia purpurea* (Lentibulariaceae): Is Carnivory Important.** *American Journal of Botany* 88: 170-176. **2001**.

Rivadavia, F., K. Kondo, M. Kato, and M. Hasebe. **Phylogeny of the Sundews, *Drosera* (Droseraceae) Based on Chloroplast rbcL and Nuclear Ribosomal DNA Sequences.** Proceedings. The 4th International Carnivorous Plant Conference. Tokyo, Japan. June 21-23. **2002**.

Rivadavia, Fernando. **Four New Species of Sundews, *Drosera* (Droseraceae), from Brazil.** *Carnivorous Plant Newsletter* 32: 79-92. **2003**.

Rivadavia, Fernando. **A *Genlisea* Myth Is Confirmed.** *Carnivorous Plant Newsletter* 36: 122. **2007**.

Roberts, Patricia R. and H. J. Oosting. **Responses of Venus Fly Trap (*Dionaea muscipula*) to Factors Involved in Its Endemism.** *Ecological Monographs* 28: 193-218. **1958**.

Robins, R. J. **The Nature of the Stimuli Causing Digestive Juice Secretion in *Dionaea muscipula* Ellis (Venus's Flytrap).** *Planta (Berl.)* 128: 263-265. **1976**.

Romanowski, Nick. *Gardening with Carnivores - Sarracenia Pitcher Plants in Cultivation & in the Wild*. Sydney: University of New South Wales Press Ltd. **2002**.

Romeo, John T., John D. Bacon and Tom J. Mabry. **Ecological Considerations of Amino Acids and Flavonoids in** *Sarracenia* **Species**. *Biochemical Systematics and Ecology* 5: 117-120. **1977**.

Rondeau, J. Hawkeye. *Carnivorous Plants of the West*. Vol. II. **1995**.

Rondeau, J. H. and Steiger, J.F. *Pinguicula macroceras* subsp. *nortensis*, **a New Subspecies of** *Pinguicula* **(Lentibulariaceae) from the California-Oregon Border**. *International Pinguicula Study Group Newsletter* 8: 3-8. **1997**.

Rondeau, Hawkeye and E. Rondeau. **The Search for** *Utricularia ochroleuca* **in Western North America**. *Carnivorous Plant Newsletter* 31: 4-8. **2002**.

Rondeau, Hawkeye. *Pinguicula* **in the Shadow of Mount Shasta, California**. *Carnivorous Plant Newsletter* 40: 50-55. **2011**.

S

Sahashi, Norio and Masa Ikuse. **Pollen Morphology of** *Aldrovanda vesiculosa* L. *Journal of Jap. Bot.* 48(12): 22-28. **1973**.

Sasago, Akira and Takao Shibaoka. **Water Extrusion in the Trap Bladders of** *Utricularia vulgaris* I. **A possible pathway of water across the bladder wall**. *Journal of Plant Research* 98(1): 55-66. **1985**.

Sasago, Akira and Takao Shibaoka. **Water Extrusion in the Trap Bladders of** *Utricularia vulgaris* II. **A possible mechanism of water outflow**. *Journal of Plant Research* 98(1): 113-124. **1985**.

Scala, J., K. Iott, D. W. Schwab and F. E. Semersky. **Digestive Secretion of** *Dionaea muscipula* **(Venus's-Flytrap)**. *Plant Physiology* 44: 367-371. **1969**.

Schlauer, Jan. **"New" Data Relating to the Evolution and Phylogeny of Some Carnivorous Plant Families**. *Carnivorous Plant Newsletter* 26: 34-38. **1997**.

Schlauer, Jan. **Fossil** *Aldrovanda* **- Additions**. *Carnivorous Plant Newsletter* 26: 98. **1997**.

Schlauer, Jan. **Carnivorous Plant Systematics**. *Carnivorous Plant Newsletter* 39: 8-24. **2010**.

Schnell, Donald. **More About the Sunshine Pitchers**. *Garden Journal (NYBG)* 24(5): 146-147. **1974**.

Schnell, Donald E. and R. Sivertsen. **A Late Summer CP Foray into Michigan and Ontario**. *Carnivorous Plant Newsletter* 3: 50-51. **1974**.

Schnell, Donald E. **The Kaleidoscope of** *Sarracenia*. *Carnivorous Plant Newsletter* 3: 7-10. **1974**.

Schnell, Donald E. and Daniel W. Krider. **Cluster Analysis of the Genus** *Sarracenia* L. **in the Southeastern United States**. *Castanea* 41: 165-176. **1976**.

Schnell, Donald E. *Carnivorous Plants of the United States and Canada*. Winston-Salem, North Carolina: John F. Blair. **1976**.

Schnell, Donald E. **Infraspecific Variation in** *Sarracenia rubra* **Walt.: Some Observations**. *Castanea* 42: 149-170. **1977**.

Schnell, Donald E. *Sarracenia flave* L.: **Infraspecific Variation in Eastern North Carolina**. *Castanea* 43: 1-20. **1978**.

Schnell, Donald E. *Sarracenia* L.: **Petal Extract Chromatography**. *Castanea* 43: 107-115. **1978**.

Schnell, Donald E. **Systematic Flower Studies of** *Sarracenia* L. *Castanea* 43: 211-220. **1978**.

Schnell, Donald E. **A Critical Review of Published Variants of** *Sarracenia purpurea* L. *Castanea* 44: 47-59. **1979**.

Schnell, Donald E. *Sarracenia rubra* **Walter ssp.** *Gulfensis*: **A New Subspecies**. *Castanea* 44: 217-223. **1979**.

Schnell, Donald. *Drosera linearis*. *Carnivorous Plant Newsletter* 9: 16-18. **1980**.

Schnell, Donald E. **A Photographic Primer of Variants of** *Sarracenia rubra* **Walt**. *Carnivorous Plant Newsletter* 11: 41-45. **1982**.

Schnell, Donald. **Special Literature Review: The Carnivorous Plants by Juniper, et al**. *Carnivorous Plant Newsletter* 18: 55-57. **1989**.

Schnell, Donald. **Special Book Review: The Genus** *Utricularia* **by Peter Taylor**. *Carnivorous Plant Newsletter* 19: 51. **1990**.

Schnell, Donald. **A Special Issue Dedication - To Peter Taylor**. *Carnivorous Plant Newsletter* 20: 4-5. **1991**.

Schnell, Donald. **Peter Taylor - A Short and Informal Biographical Sketch**. *Carnivorous Plant Newsletter* 20: 6-7. **1991**.

Schnell, Donald. **Book Review: Venus's Flytrap - Aphrodite's Mousetrap**. *Carnivorous Plant Newsletter* 20: 124-125. **1991**.

Schnell, Donald. *Drosera anglica* **Huds. vs.** *Drosera x anglica*: **What Is the Difference?** *Carnivorous Plant Newsletter* 28: 107-115. **1999**.

Schnell, Donald E. *Carnivorous Plants of the United States and Canada*. 2nd ed. Portland, Oregon: Timber Press. **2002**.

Shanos, Gregory T. **Action Potentials in the Venus Flytrap**. *Carnivorous Plant Newsletter* 15: 16-17,26. **1986**.

Sheridan, Philip M. **What is the Identity of the West Gulf Coast Pitcher Plant,** *Sarracenia alata* **Wood?**. *Carnivorous Plant Newsletter* 20: 102-110. **1991**.

Shimizu, Kiyoshi. *Insectivorous Plants Photography*. Tokyo: Seibundo-Shinkosha. **1966**.

Shreve, Forrest. **The Development and Anatomy of** *Sarracenia purpurea*. *Botanical Gazette* 42: 107-126. **1906**.

Simons, Paul. **How Exclusive Are Carnivorous Plants?** *Carnivorous Plant Newsletter* 10: 65-68,79-80. **1981**.

Slack, Adrian. *Carnivorous Plants*. Cambridge, Massachusetts: MIT Press. **1979**.

Slack, Adrian. *Insect-Eating Plants and How To Grow Them*. Sherborne, Dorset, England: Alphabooks. **1986**.

Snyder, Ivan. **Colchicine Treatment on Sterile Hybrid Sundews**. *Carnivorous Plant Newsletter* 29: 4-10. **2000**.

Sorenson, Daniel R. and William T. Jackson. **The Utilization of Paramecia by the Carnivorous Plant** *Utricularia gibba*. *Planta (Berl.)* 83: 166-170. **1968**.

Speirs, D. C. **The Evolution of Carnivorous Plants**. *Carnivorous Plant Newsletter* 10: 62-65. **1981**.

Stafleu, A. Frans. **Historiae Naturalis Classica: Wilkes, de Candolle, Weddell**. *Taxon* 21(1): 167-176. **1972**.

Stromberg-Wilkins, Juliet C. **Autecological Studies of** *Drosera linearis*, **a Threatened Sundew Species**. *The University of Wisconsin-Milwaukee, Field Station Bulletin* 17(1): 1-16. **1984**.

Studnicka, Miloslav. **Observations on Two Different Forms of** *Utricularia reniformis*. *Carnivorous Plant Newsletter* 33: 47-51. **2004**.

Stuhlman, Otto, Jr. **A Physical Analysis of the Opening and Closing Movements of the Lobes of Venus' Fly-Trap**. *Bulletin of the Torrey Botanical Club* 75: 22-44. **1948**.

Subramanyam, K and V. Abraham. **Studies on the Traps of Some Indian Species of** *Utricularia* **L**. *Bulletin of Botanical Survey of India* 9: 201-205. **1967**.

Swales, Dorothy E. *Sarracenia purpurea* **L. As Host and Carnivore at Lac Carre, Terrebonne Co., Quebec**. *Le Naturaliste Canadien* 96: 759-763. **1969**.

Swales, Dorothy E. *Sarracenia purpurea* **L. As Host and Carnivore at Lac Carre, Terrebonne Co., Quebec. Part II**. *Quebec Le Naturaliste Canadien* 99: 41-47. **1972**.

Sydenham, P. H. and G. P. Findlay. **The Rapid Movement of the Bladder of** *Utricularia* **sp**. *Aust. J. Biol.Sci.* 26: 1115-1126. **1973**.

Sydenham, P. H. and G. P. Findlay. **Transport of Solutes and Water by Resetting Bladders of** *Utricularia*. *Australian Journal of Plant Physiology* 2: 335-351. **1975**.

T

Taylor, Peter. **A New Combination in** *Genlisea* **(Lentibulariaceae)** *Genlisea hispidula*. *Kew Bulletin* 26(3): 444. **1972**.

Taylor, Peter. *The Genus Utricularia*. London: Royal Botanic Gardens, Kew. **1989**.

Taylor, Peter. **The Genus** *Genlisea*. *Carnivorous Plant Newsletter* 20: 20-26. **1991**.

Taylor, Peter. *Utricularia* **in North America North of Mexico**. *Carnivorous Plant Newsletter* 20: 8-20, 36-58. **1991**.

Taylor, Peter. **The Genus** *Genlisea* **St. Hil. An Annoted Bibliography**. *Carnivorous Plant Newsletter* 20: 27-35, 59. **1991**.

Thurston, E. Laurence and Frank Seabury. **A Scanning Electron Microscopic Study of the Utricle Trichomes in** *Utricularia biflora* **LAM**. *Botanical Gazette* 136(1): 87-93. **1975**.

Tokes, Zoltan A., Wang Cheo Woon and Susan M. Chambers. **Digestive Enzymes Secreted by the Carnivorous Plant** *Nepenthes macferlanei* **L**. *Planta (Berl.)* 119: 39-46. **1974**.

U

Ueda, Minoru, Takashi Tokunaga, Tsukasa Katanami, Noboru Takada, Rie Suzuki and Katsuhiko Kondo. **Chemical Substance Concerning the Leaf-Movement of** *Dionaea muscipula*. The 4th International Carnivorous Plant Conference, Tokyo, Japan. Proceedings: 97-99. **2002**.

Ueda, Minoru. **Chemical Substance Concerning the Venus Flytrap Leaf Movement**. *Iden (heritage). Science of Living Matter*. Volume 59. No.4: 55-60. **2005**.

Ueda, Minoru, Takashi Tokunaga, Masahiro Okada, Yoko Nakamura, Noboru Takada, Rie Suzuki and Katsuhiko Kondo. **Trap-Closing Chemical Factors of the Venus Flytrap (***Dionaea muscipula* **Ellis)**. *ChemBioChem* 11: 2378-2383. **2010**.

V

Vincent, O., Roditchev, I., and Marmottant, P. **Spontaneous Firings of Carnivorous Aquatic** *Utricularia* **Traps: Temporal Patterns and Mechanical Oscillations**. PLoS ONE 6: e20205. **2011**.

Vincent, O., and Marmottant, P. **Carnivorous** *Utricularia*: **The Buckling Scenario**. *Plant Signal. Behav.* 6: 1752-1754. **2011**.

W

Wan, A. S., R. T. Aexel, R. B. Ramsey and H. J. Nicholas. **Nepenthaceae - Sterols and Triterpenes of the Pitcher Plant**. *Phytochemical Reports* 10: 456-461. **1971**.

Wherry, Edgar T. **The Geographic Relations of** *Sarracenia purpurea*. *Bartonia. Proceedings of the Philadelphia Botanical Club*. 1-6. **1933**.

Wherry, Edgar T. **The Appalachian Relative of** *Sarracenia flava*. *Bartonia. Proceedings of the Philadelphia Botanical Club*. 7-8. **1933**.

Wherry, Edgar T. **Notes on** *Sarracenia* **subspecies**. *Castanea* 37(2): 146-147. **1972**.

Wherry, Edgar T. **Reminiscences on Carnivorous Plants**. *Carnivorous Plant Newsletter* 2: 35-37. **1973**.

Wherry, E. T. **On** *Darlingtonia* **vs.** *Chrysamphora*. *Carnivorous Plant Newsletter* 3: 22. **1974**.

Williams, Stephen E. and Barbara G. Pickard. **Properties of Action Potentials in** *Drosera* **Tentacles**. *Planta (Berl.)* 103: 222-240. **1972**.

Williams, Stephen E. and Barbara G. Pickard. **Connections and Barriers Between Cells of** *Drosera* **Tentacles in Relation to Their Electrophysiology**. *Planta (Berl.)* 116: 1-16. **1974**.

Williams, Stephen E. **How Venus' Flytraps Catch Spiders and Ants**. *Carnivorous Plant Newsletter* 9: 65,75-78. **1980**.

Williams, Stephen E. **How Venus' Flytraps Catch Spiders and Ants**. *Carnivorous Plant Newsletter* 9: 91,100. **1980**.

Williams, Stephen E. and A. B. Bonnett. **Leaf Closure in the Venus Flytrap: An Acid Growth Response**. *Science* 218: 1120-1122. **1982**.

Williams, Stephen E. **Mechanisms of Trap Movement 1: Rapid Growth in *Drosera*, *Dionaea* and Scientific Notions of How Venus's Flytrap Close**. *Carnivorous Plant Newsletter* 21: 14-17. **1992**.

Williams, Stephen E. **Mechanisms of Trap Movement II: Does *Aldrovanda* Close by a Turgor Mechanism? A Question of How Much, Where, and When**. *Carnivorous Plant Newsletter* 21: 46-51. **1992**.

Williams, Stephen E. **What Are the Nearest Noncarnivorous Relatives of Carnivorous Plants?** *Carnivorous Plant Newsletter* 22: 31-33. **1993**.

Willis, J. H. **Historical Notes on the W. A. Pitcher Plant, *Cephalotus follicularis*.** *The Western Australian Naturalist* 10: 1-7. **1965**.

Wood, Carroll E., Jr. **Evidence of the Hybrid Origin of *Drosera anglica*.** *Rhodora.* 57(676): 105-130. **1955**.

Wood, C. E. Jr. and R. K. Godfrey. ***Pinguicula* (Lentibulariaceae) in the Southeastern United States**. *Rhodora* 59: 217-230. **1957**.

Wood, Carroll E., Jr. **The Genera of Sarraceniaceae and Droseraceae in the Southeastern United States**. *Journal of the Arnold Arboretum* 41: 152-163. **1960**.

Wood, Carroll E., Jr. **On the Identity of *Drosera brevifolia*.** *Journal of the Arnold Arboretum.* 47(2): 89-99. **1966**.

Wynne, Frances E. ***Drosera* in Eastern North America**. *Bulletin of the Torrey Botanical Club.* 71(2): 166-174. **1944**.

Y

Yoshimaru, Katsuyuki. **Studies on the Chromosome Number and Karyotype of *Pinguicula ramosa* Miyoshi (C)**. Journal of Jap. Bot. 48(10): 289-294. **1973**.

Index

- A -
Aldrovanda vesiculosa 18, 36, 40
- B -
bladderwort > SEE *Utricularia*
Brocchinia 15, 18, 37
  hectioides 37
  reducta 37
Bromeliaceae 37
butterwort > SEE *Pinguicula*
Byblidaceae 37
Byblis 18, 28, 37
- C -
California pitcher plant > SEE
  Darlingtonia
Calopogon tuberosus 22, 91
canebrake pitcher plant > SEE
  Sarracenia rubra ssp. *alabamensis*
Caryophyllales 15, 20, 37
Catopsis berteroniana 37
Cephalotus follicularis 36, 37
Cleistes bifaria 58
cobra lily > SEE *Darlingtonia*
cobra plant > SEE *Darlingtonia*
corkscrew plant > SEE *Genlisea*
- D -
Darlingtonia californica 20, 26, 33,
  94-127
  form *viridiflora* 99
devil's claw 36
Dionaea muscipula 18, 23, **166-185**
Drosera **128-165**
  anglica 130, **150-153**
  brevifolia **158-159**
  burmannii 40, 138
  capillaris **156-157**
  filiformis 128, 130, 137, **160-165**
    variety *filiformis* 162-163
    variety *tracyi* 160-161, 164-165
  glanduligera 40, 138
  intermedia 129, 139, **154-155**
  linearis 131, **144-149**
  rotundifolia 132, 135, **140-143**
Droseraceae 37, **129**, **167**
Drosophyllaceae 37
Drosophyllum 18, 26, 36
dwarf huckleberry 63
- E -
Ericales 37
- G -
Gaylussacia dumosa 63
Genlisea 36
green pitcher plant > SEE *Sarracenia*
  oreophila
- H -
Heliamphora 18, 37, 88, 95
hooded pitcher plant > SEE *Sarracenia*
  minor

- I -
Ibicella 37
  lutea 37
  parodii 37
- L -
Lamiales 37
Lentibulariaceae 37, **187**, **223**
linear-leaved sundew > SEE *Drosera*
  filiformis
- M -
marsh pitcher plant > SEE *Heliamphora*
Martyniaceae 37
- N -
Nepenthaceae 37
Nepenthes 18, 36, 37
- O -
Oxalidales 36
- P -
Paepalanthus bromelioides 37
parrot pitcher plant > SEE *Sarracenia*
  psittacina
Philcoxia 37
Pinguicula 37, 94, **186-221**
  alpine 185
  caerulea **208-209**
  ionantha 38, 190, 191, **216-217**
  lutea 188, **204-207**
  macroceras 16, 195, **200-203**
  planifolia **212-215**
  primuliflora **218-221**
  pumila **210-211**
  vulgaris 189, **198-199**
pitcher plant > SEE *Sarracenia*
Poales 36
Pogonia ophioglossoides 59
Portuguese dewy pine > SEE
  Drosophyllum
Proboscidea 37
  lousianica 37
  parviflora 37
- R -
rainbow plant > SEE *Byblis*
Roridula 36
Roridulaceae 37
round-leaved sundew > SEE *Drosera*
  rotundifolia
- S -
Sarracenia **42-93**
  alata 44, **54-57**
  flava 50, 51, **74-79**
    variety *atropurpurea* 74
    variety *cuprea* 74
    variety *flava* 74
    variety *maxima* 77
    variety *ornata* 79
    variety *rubricorpora* 74
    variety *rugelii* 75, 78

  leucophylla 42, 43, 47, 53, **70-73**
    variety *alba* 68
  minor **80-83**
  oreophila 49, **66-69**
    variety *ornata* 66
  psittacina **84-87**
  purpurea **88-93**
    subspecies *purpurea* 24-25, **90-93**
    subspecies *venosa* **88-89**
      variety *burkii* 88
      variety *montana* 88
      variety *venosa* 88
  rubra 58-65
    subspecies *alabamensis* **62-65**
    subspecies *gulfensis* **58-59**
    subspecies *jonesii* 62
    subspecies *rubra* 58
    subspecies *wherryi* **60-61**
Sarraceniaceae 37, **43**, **95**
sundew > SEE *Drosera*
- T -
thread-leaf sundew > SEE *Drosera*
  filiformis
Triphyophyllum peltatum 18, 32, 36
tropical pitcher plant > SEE *Nepenthes*
- U -
Utricularia 33, 36, 91, 142, **222-262**
  biflora 236
  cornuta 34, 222, **260-261**
  fibrosa 237
  floridana **238-241**
  geminiscapa 229, **250-251**
  gibba 226, **236**
  inflata 231, 234, **252-253**
  intermedia 224, 229, **244-247**
  juncea 261
  macrorhiza **242-243**
  minor **248-249**
  purpurea **256-258**
  radiate **254-255**
  striata 237
  subulata **258-259**
  vulgaris 242
- V -
Venus flytrap > SEE *Dionaea muscipula*
- W -
waterwheel plant > SEE *Aldrovanda*
  vesiculosa
Western Australian pitcher plant > SEE
  Cephalotus follicularis
- Y -
yellow trumpet > SEE *Sarracenia flava*

A photograph is a capture of a moment —
the photographer is there to witness the event,
creating an image frozen in time forever.

On one weekend in May, 2012, I visited Butterfly Valley in northern California. I have visited this nature preserve a number of times in the past, and I have many pictures of both flowers and foliage of the cobra plants growing there ... Well, this time I may be lucky to see their mysterious pollinators! Butterfly Valley, after all, is *mere* 8-hour drive from my home in southern California. When I arrived there, it was cloudy. The plants were blooming, but it was not a great day for a photo shoot. After poking around in the seep, I stayed in a local motel. Next day, before heading home, I entered the preserve one more time. The morning was foggy, as typical in this area, but the sky remained dark and a drizzle persisted from the previous night. I walked around, but not much excitement. Time to go ... then, suddenly, the morning sun pierced through the clouds, lighting up the montane theatre. I hastily composed a picture — with colorful blossoms shining brightly against the grey sky. The light did not last long, but I got my shot!

Made in the USA
Monee, IL
01 May 2024

c5c56987-ffc2-484b-ab25-8dfea7371278R01